Nanogenerators

Edited by Sang Jae Kim,
Arunkumar Chandrasekhar
and Nagamalleswara Rao Alluri

Published in London, United Kingdom

IntechOpen

Supporting open minds since 2005

Nanogenerators
http://dx.doi.org/10.5772/intechopen.78915
Edited by Sang Jae Kim, Arunkumar Chandrasekhar and Nagamalleswara Rao Alluri

Contributors
Panpan Li, Jeongjae Ryu, Seungbum Hong, Andrei Kholkin, Svitlana Kopyl, Vladislav Slabov, Marco P. Soares Dos Santos, Seongmun Eom, Chi Hao Liow, Sang Jae Kim, Arunkumar Chandrasekhar, Nagamalleswara Rao Alluri, Venkateswaran Vivekananthan, Yuvasree Purusothaman, Gaurav Khandelwal, Prateek Asthana, Gargi Khanna, Zhen Wen

First published in London, United Kingdom, 2020 by IntechOpen
IntechOpen is the global imprint of INTECHOPEN LIMITED, registered in England and Wales, registration number: 11086078, 7th floor, 10 Lower Thames Street, London, EC3R 6AF, United Kingdom
Printed in Croatia

British Library Cataloguing-in-Publication Data
A catalogue record for this book is available from the British Library

Additional hard and PDF copies can be obtained from orders@intechopen.com

Nanogenerators
Edited by Sang Jae Kim, Arunkumar Chandrasekhar and Nagamalleswara Rao Alluri
p. cm.
Print ISBN 978-1-83881-059-7
Online ISBN 978-1-83881-060-3
eBook (PDF) ISBN 978-1-83881-061-0

For EU product safety concerns:
IN TECH d.o.o., Prolaz Marije Krucifikse Kozulić 3, 51000 Rijeka, Croatia, info@intechopen.com or visit our website at intechopen.com.

Meet the editors

Sang Jae Kim, PhD, is a professor in the Department of Mechatronics Engineering and the Department of Advanced Convergence Technology and Science at Jeju National University (JNU), South Korea. He received his PhD in Electrical Communication Engineering from Tohoku University, Japan. He was a visiting research scholar in the Department of Materials Science at the University of Cambridge, UK, and Georgia Institute of Technology, USA, as well as a senior researcher at the National Institute of Materials Science. He has published more than 150 research articles in well-reputed international journals. His research disciplines include nanomaterials for energy and electronics applications, Josephson devices, microelectromechanical systems, supercapacitors, nanogenerators, and nano-biosensors.

Arunkumar Chandrasekhar, PhD, is currently Assistant Professor in the Department of Sensors and Biomedical Technology, Vellore Institute of Technology, India. He worked as a postdoctoral researcher at the Nanomaterials and Systems Laboratory, South Korea. He received his PhD in Mechatronics Engineering from Jeju National University, South Korea, where he was a recipient of a scholarship from the Korean Government Scholarship Program. He received the prestigious Brain Korea 21+ Business Research Award from the Ministry of Education for excellence in research work. He has published forty-three research articles, six journal cover pages, and two book chapters in reputed international journals. He is interested in wearable triboelectric nanogenerators, battery-free electronic devices, energy storage devices, microelectromechanical systems, and self-powered devices.

Nagamalleswara Rao Alluri, PhD, is currently a postdoctoral researcher at the Nanomaterials and System Laboratory, Department of Mechatronics Engineering, Jeju National University (JNU), South Korea. He received the young investigator project from the National Research Foundation of Korea as a principal investigator. He received the prestigious BK21+ Business Research Award from the Ministry of Education, and the Presidential Award from JNU for excellence in research work (2014–2018). Dr. Alluri has published forty-five research articles, two book chapters, and five journal cover pages in well-reputed international journals. He received a PhD in Mechanical Engineering from JNU; MTech in Sensor System Technology from Vellore Institute of Technology, India; and MSc in Condensed Matter Physics from Andhra University, Visakhapatnam, India. His research interest is functional nanomaterials for energy and sensor applications.

Contents

Preface **XI**

Section 1
Triboelectric Nanogenerator **1**

Chapter 1 **3**
Triboelectric Nanogenerators: Design, Fabrication, Energy Harvesting, and Portable-Wearable Applications
by Venkateswaran Vivekananthan, Arunkumar Chandrasekhar, Nagamalleswara Rao Alluri, Yuvasree Purusothaman, Gaurav Khandelwal and Sang-Jae Kim

Chapter 2 **21**
Design of Electrode Materials for Stretchable Triboelectric Nanogenerators
by Zhen Wen

Section 2
Piezoelectric Nanogenerator **37**

Chapter 3 **39**
Development of Vibration Piezoelectric Harvesters by the Optimum Design of Cantilever Structures
by Prateek Asthana and Gargi Khanna

Chapter 4 **51**
Ferroelectric Polymer PVDF-Based Nanogenerator
by Jeongjae Ryu, Seongmun Eom, Panpan Li, Chi Hao Liow and Seungbum Hong

Chapter 5 **73**
Piezoelectricity in Self-Assembled Peptides: A New Way towards Electricity Generation at Nanoscale
by Vladislav Slabov, Svitlana Kopyl, Marco P. Soares dos Santos and Andrei Kholkin

Section 3
Nanogenerator for Biomedical Applications **91**

Chapter 6 **93**
Piezoelectric/Triboelectric Nanogenerators for Biomedical Applications
by Panpan Li, Jeongjae Ryu and Seungbum Hong

Preface

This book introduces technological developments in the field of energy conversion. Currently, various energy conversion technologies are available, such as photovoltaic cells, electromagnetic generators, and so on. This book introduces a new energy conversion device called a "nanogenerator." It focuses on two types of nanogenerators: triboelectric and piezoelectric. Chapters cover fundamental topics such as material identification, synthesis processes, device fabrication, and self-powered applications. Detailed results from research in the field are also provided.

Moreover, the book discusses the concept of wearable, portable, and implantable devices and their importance in our daily lives. Stretchable electrodes play a significant role in wearable devices and this book examines the design of electrode materials for stretchable triboelectric nanogenerators. The information contained herein will help researchers to generate more ideas and techniques for triboelectric and piezoelectric nanogenerators as well as future inventions.

Sang Jae Kim and Nagamalleswara Rao Alluri
Jeju National University,
South Korea

Arunkumar Chandrasekhar
Vellore Institute of Technology University,
India

Section 1

Triboelectric Nanogenerator

Chapter 1

Triboelectric Nanogenerators: Design, Fabrication, Energy Harvesting, and Portable-Wearable Applications

Venkateswaran Vivekananthan, Arunkumar Chandrasekhar, Nagamalleswara Rao Alluri, Yuvasree Purusothaman, Gaurav Khandelwal and Sang-Jae Kim

Abstract

Scavenging energy from our day-to-day activity into useful electrical energy be the best solution to solve the energy crisis. This concept entirely reduces the usage of batteries, which have a complex issue in recycling and disposal. For electrical harvesting energy from vibration energy, there are few energy harvesters available, but the fabrication, implementation, and maintenances are quite complicated. Triboelectric nanogenerators (TENG) having the advantage of accessible design, less fabrication cost, and high energy efficiency can replace the battery in low-power electronic devices. TENGs can operate in various working modes such as contact-separation mode, sliding mode, single-electrode mode, and free-standing mode. The design of TENGs with the respective operating modes employed in generating electric power as well as can be utilized as a portable and wearable power source. The fabrication of triboelectric layers with micro-roughness could enhance the triboelectric charge generation. The objective of this chapter is to deal with the design of triboelectric layers, creating micro structured roughness using the soft-lithographic technique, fabrication of TENGs using different working modes, energy harvesting performance analysis, powering up commercial devices (LEDs, displays, and capacitors), and portable-wearable applications.

Keywords: triboelectric nanogenerators, micro-roughness, soft-lithography, TENG working modes, portable-wearable

1. Introduction

In the last few decades, the boom in consumer electronics rapidly creates the need for a sustainable portable and wearable power source. There are too many low-power electronic devices that are introduced to enhance the quality of our life, such as sensors, communication devices, GPS devices, implantable, and health monitoring devices [1–4]. All these electronic gadgets require electric power in the range of few microwatts to milliwatts. With the invention of batteries, these devices can use in various portable applications. However, the drawback in batteries limits the purpose and be an obstacle to miniaturize the sensors. As per the batteries in concern, the batteries require periodical replacement, and the recycling of batteries is still a significant challenge. The precursor used for the fabrication of the batteries is highly hazards to human life and creates pollution to the environment [5]. The disposal has many concerns regarding soil and water contamination as they contain a lot of heavy metals. The future of electronics depends on the portable and wearable sensors such as flexible electronics [6, 7], e-skin-based sensor devices [8], and implantable bio-medical electronic systems [9]. Operating the low-power electronic devices, a sustainable and highly reliable power source is required. The power sources should scavenge power from waste mechanical energy. The majority of energy that we consume today is from many nonrenewable sources such as thermal power plants that operate on coal, and diesel power plants operating with the help of oil. These natural resources are limited in supply, and it takes thousands of years to regenerate. The usage of fossil fuels such as coal and oil in the generation of electricity creates massive pollution, and it leads to the reduction of life span to humans. To overcome this issue, an alternative energy harvesting approach had been introduced across the globe to protect humanity and overcome the growing energy crisis.

The conventional approach to generate electrical energy is by using an electromagnetic generator (EMG). These generators harvest effectively under the high frequency of mechanical energy inputs [10]. However, there is an abundant low-frequency mechanical energy around the globe as well as from various sources such as vibration, wind, water wave, human motions, and vehicle motions. These low-frequency motions cannot be harvested effectively using an EMG. The solution to this problem addressed in 2006 with the invention of nanogenerators (NGs) [11]. The NGs with various classifications and effects such as piezoelectric nanogenerator (PNG) [12–15], triboelectric nanogenerator (TENG) [16], thermoelectric generator (TEG) [17], and pyroelectric generator (PYG) [18] had been introduced in the past decades for a variety of energy harvesting and self-powered applications.

Furthermore, the energy harvesters are performing some real-time applications by integrating it with battery or other storage devices. Apart from energy harvesting, these NG can work as a self-powered sensors such as pH sensors [19], glucose sensors [20], chemical sensors [21], humidity sensors [22, 23], temperature sensors [24], air pressure sensors [25], motion sensor [26], optical sensor [27, 28], and strain sensors [29]. To enhance the output power of the TENG devices, researchers across the globe performed the hybridization of these devices and studied the possibility of its use in extensive applications. To be used for a wide range of applications, TENG-PENG [30] hybrid had been reported with high instantaneous power density. Similarly, there are many reports for scavenging the water wave energy by TENG-EMG hybrid configuration [31]. The generated output is much higher and can use for GPS tracking and positioning system. A similar concept is demonstrating by using magnetic particles instead of magnets in the EMG component [32]. The motion of particles activates the triboelectric part of the hybrid device. The issues that TENGs have faced so far are its bio-compatibility, performance reduction concerning humidity. The problems rectified by introducing a biodegradable, edible TENG [33], and a fully packed water-resistant TENG. The fully packed TENG [34] can protect the device from humidity issues and maintains its performance for a more extended period. The further advancements in NGs lead to the design of flexible electronic-skin-based devices for sensing various parameters, microfluidic-based devices [35], and other portable-wearable devices making NGs a sustainable power source. This chapter covers the basics of TENG, fundamental working modes of TENG and fabrication, and energy harvesting and its application in portable and wearable technologies.

2. Working modes of TENG

The mechanism behind TENG is triboelectrification and electrostatic induction effects; with the rapid advancement in the field of nanogenerators and TENG, the exploration leads to the investigation of fundamental working modes of TENG, which covers four basic modes. The modes include a vertical contact separation mode, linear sliding mode, single-electrode mode, and free-standing mode. These modes require two different triboelectric materials with proper electrode connections with proper insulation between each layer. The combination may be either dielectric-dielectric or metal-dielectric arrangements. The basic principle behind all the four modes is that whenever there is a displacement in any of the triboelectric layers, the electrostatic charge movements break the electrostatic status present, leading to the development of potential difference between the electrodes. In the repeated mechanical actuation of layers with forward and reverse direction makes the triboelectric layer to generate forward and reverse potential between the electrodes, making the positive and negative peaks in the TENG output generating the AC signal. The four different working modes are shown schematically in **Figure 1**.

2.1 Contact separation mode

In the contact separation mode TENG, the triboelectrification occurs by the contact and separation process of two triboelectric materials or layers, as shown in **Figure 1a**. The process may either happen between two different dielectric materials or may be a dielectric and metallic layers. This model has a significant advantage with its simple design, easy fabrication, and low cost. This mode of TENG is also the first developed and demonstrated for powering low-power electronics. This mode of TENG can also make as a multi-unit stacking for the enhanced output performance.

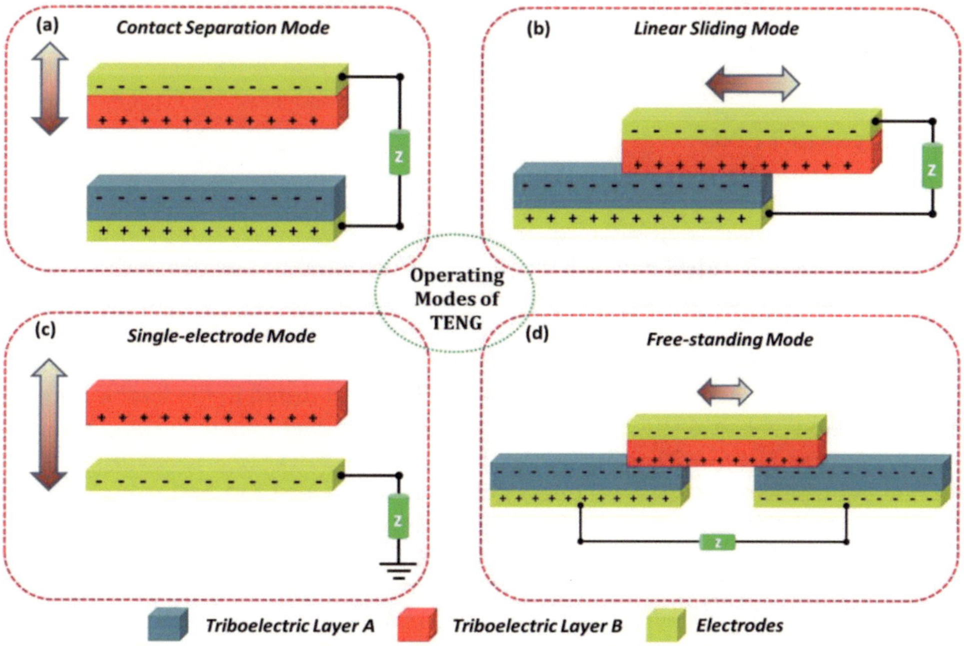

Figure 1.
Schematic representation of the four fundamental operating modes of TENG.

2.2 Linear sliding mode

In the linear sliding mode, the charge generation is by the relative to and from sliding between the layers of TENG. The construction is almost similar to the contact separation mode with electrodes attached to the back of the triboelectric layers, but the displacement is in sideward, as shown in **Figure 1b**. The sliding mode has a low figure-of-merits compared with the vertical contact separation mode due to its protracted displacement in the sliding process. The advantage of this mode is that it can generate more charge density with a highly effective charge generation due to its high contact area. Also, by introducing more grating structures, the output performance can be enhanced. The sliding mode TENG can also be able to operate rotationally with cylindrical grating structures.

2.3 Single-electrode mode

The simplest structure of TENG is the single-electrode mode TENG, but the output performance is too low due to the small charge transfer, making the generated voltage and current to be less, but it is highly suitable for self-powered applications. The advantage of this device is that it overcomes the application limitation due to the contact wire obstructing at both sides in the contact separation mode and sliding mode TENG devices. The basic arrangement of a single-electrode mode TENG is shown in **Figure 1c**.

2.4 Free-standing mode

The free-standing model has one electrode move freely between the two electrodes or triboelectric layers. The electrodes are in a static position, and a triboelectric layer without electrode can move over it. This mode of TENG device has a high figure-of-merits and has demonstrated high output efficiency and electrical output. Also, this type of device can be fabricated easily and integrated into various real-time applications. The necessary arrangement of a single-electrode mode TENG is shown in **Figure 1d**.

3. Progress in TENG

After the invention of TENG in 2012, the field becomes an extensive investigated field with a vast number of researchers working and improving it until today [36]. The first developed TENG device is a contact separation mode device with kapton and polyester as the active layers [7]. Further investigation TENG leads to the invention of other modes in 2013 [37]. The research trend on NGs started moving forward from energy harvesting to self-powered sensing and actuation. The transparent and flexible nanogenerators were fabricated and tested in the later stage with the application in e-skin and tactile sensing. The researchers' then working towards hybridizing the TENG with other energy harvesters such as PENG and EMG for scavenging wind and water wave energy efficiently and also had performed various real-time applications connected to environmental monitoring. Very recently, by adopting these explored TENG design and operation, liquid–solid interface TENGs and liquid–liquid TENGs had introduced for scavenging water droplet energy [38]. There had been a problem in TENGs and its electrical output concerning external factors such as temperature and humidity. Recently, the humidity issues got rectified by introducing a fully packed, waterproof TENG device by our group in 2019. In addition to biodegradable TENGs, an edible TENG made of regular food items

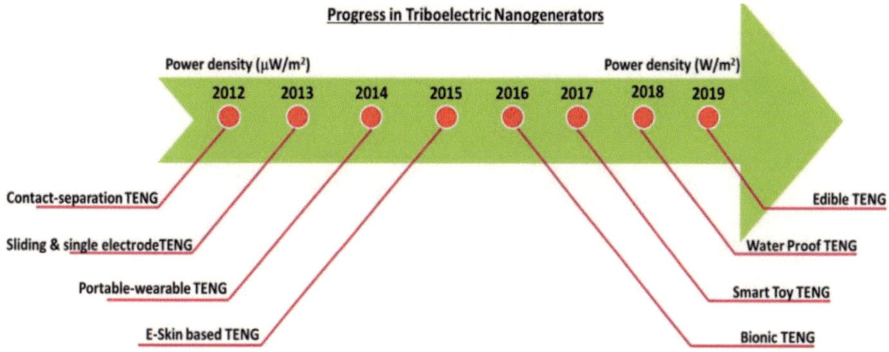

Figure 2.
Design and development of TENGs from its origin to date, with various device structures, applications, and performances.

was designed and fabricated by our group and demonstrated powering low-power electronic devices. Also, the first developed TENG device has a power density in the range of micro watts, and today, the device had reached the power density in watts. The increasing power density shows that the TENGs are reliable energy harvester and self-powered sensors which can able to commercialize shortly and to have a high possibility to replace batteries. **Figure 2** shows the road map, explaining the advancement of TENGs for various applications.

4. Fabrication of TENG device

The fabrication of a TENG device requires precise handling of the arrangement of triboelectric layers and electrodes. The improvement in the design of the TENG device is improving day-by-day. The triboelectric layer with a high surface roughness delivers high electrical output. The phenomenon is that the rough surfaces increase the contact points of the TENG device. Each micro/nano scale roughness interacts with the other triboelectric layer during its mechanical actuation. Therefore, the contact point multiplies the generated output as compared with the flat surface where the contact point is single. There are various approaches to creating surface roughness of the triboelectric layers, such as inductively coupled plasma etching, wet etching, lithographic technique, and casting. These techniques involve various advantages as well as disadvantages. The high-end lithographic and plasma etching techniques involve high-cost instruments and skilled operator, which is a disadvantage. There are various reports on simply creating surface roughness using cost-effective techniques such as using sandpaper as a mold. The sandpaper with more roughness as a mold for fabricating roughness created triboelectric layer, and the polymers such as PDMS and other silicon elastomer-based polymers poured on it, and with heat treatment, the polymer solution is then cured to form a polymeric film with microscale surface roughness. The other cost-effective technique reported is the surface modification of a used acrylonitrile butadiene styrene (ABS) polymer petri dish. The surface modification is done through wet etching using acetone as an etchant. The technique involves less time and a simple preparation process.

The ABS polymer petri dish is first washed with water and dipped in a beaker full of acetone for 90 s, as shown in **Figure 3a**, and the petri dish dried in an air atmosphere at room temperature. By this time, acetone etches the surface of the petri dish, creating micro-roughness on its surface, as shown in **Figure 3b**. The PDMS polymer is poured into the petri dish and cured in a hot air oven to get a polymeric film.

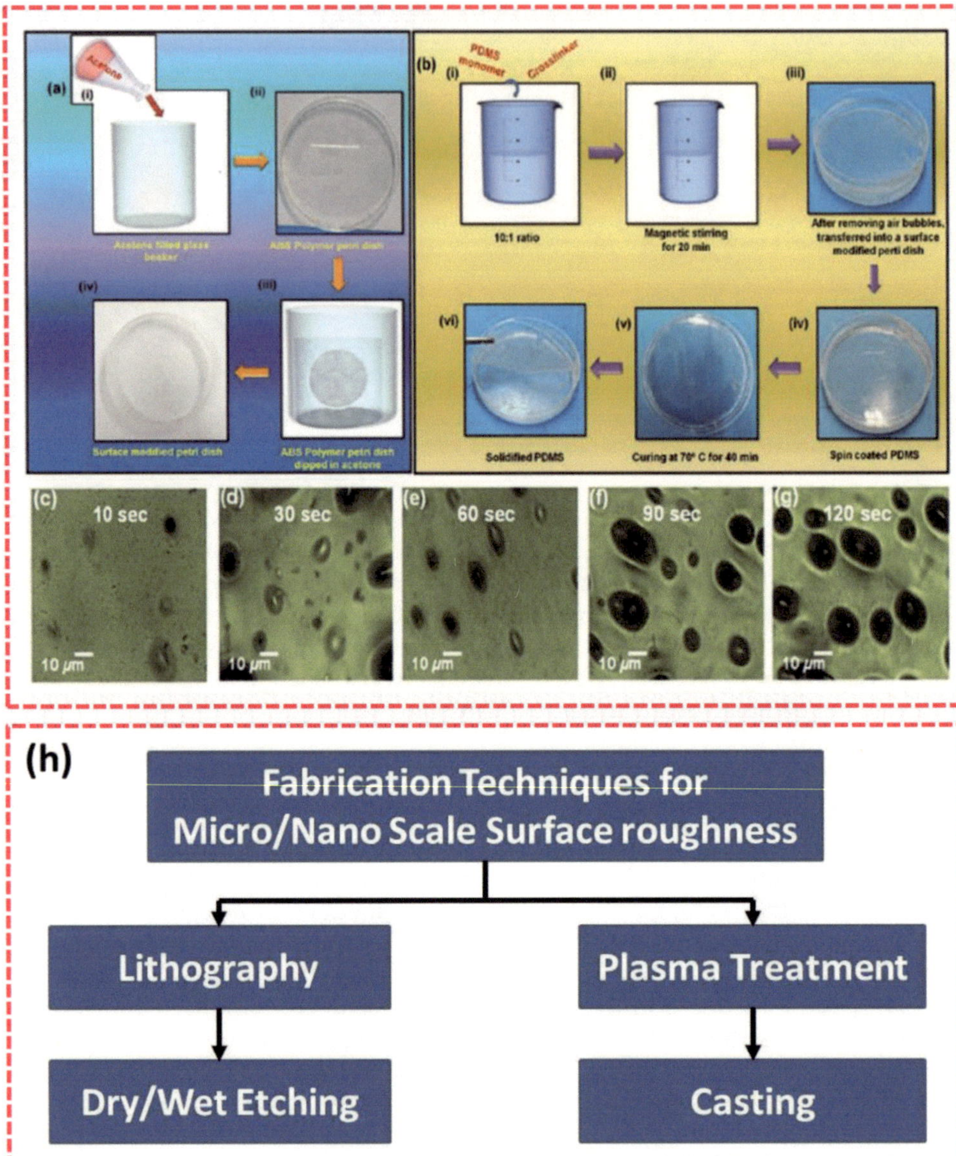

Figure 3.
(a, b) Wet etching of an ABS polymer petri dish using acetone as an etchant and fabrication of mold. The PDMS monomer is then poured on the surface-etched mold and cured to form a roughness patterned PDMS film. (c–g) FE-SEM morphology showing the surface roughness of PDMS film cured in a petri dish with different acetone treated time and their corresponding surface roughness. (h) Fabrication techniques for micro/nanoscale surface roughness [39].

Figure 3c–g shows the FE-SEM morphology of the surface of the PDMS polymer film made using roughness created petri dish mold. The petri dish with acetone treatment for a period of 90 s shows good roughness. **Figure 3h** shows the fabrication techniques involved in the fabrication of different micro or nanoscale surface roughness.

5. TENGs as a biomechanical energy harvester

The primary purpose of a TENG device is that it can harvest mechanical energy from the environment. The mechanical energy can be in any form ranging from vibration, vehicle motion, and human body motion. To scavenge the biomechanical energy and use effectively for a variety of applications, an H-TENG and SS-TENG devices report the fabrication of wearable TENG and its performance to applications. **Figure 4** shows the H-TENG device and its electrical performances with the application of an emergency exit system. The H-TENG device made up of polymeric materials by collecting the polymeric waste materials from the house garbage bin, as shown in **Figure 4a**. The positive material is cooking aluminum foil, and the harmful triboelectric materials are random plastics, and the spacer material is the utensil cleaning sponge. The entire TENG device made of household waste shows its easy fabrication and competitive cost development process. The layer-by-layer schematic of H-TENG is shown in **Figure 4b**. The statistics of plastic wastes generated and creating adverse effect to the environment is shown in **Figure 4c**. The device works on vertical contact separation mode, as shown in **Figure 4d**. The device successfully used for scavenging waste mechanical energy from the human finger, palm, hand, and leg motions, as shown in **Figure 4e** with their corresponding electrical response. The electrical output of the H-TENG device made of various polymers was measured in the laboratory using a linear motor, and their corresponding voltage and current responses are shown in **Figure 4f**. The devices were constantly powering at a maximum of 20 V/300 nA response. At last, the fabricated H-TENG device was used for an emergency exit LED lighting system, indicating the exit direction, as shown in **Figure 4g**.

In a similar approach, an SS-TENG device fabricated for scavenging biomechanical energy by placing it on the seating chair. The whole process of SS-TENG from the schematic, electrical response and applications are shown in **Figure 5**. The device consists of interdigitated electrodes (IDE) made of aluminum is attached to the bottom side of a Kapton film. The kapton film acts as a negative triboelectric layer, and the contact materials were bus cards, polyethylene cover, jeans cloth,

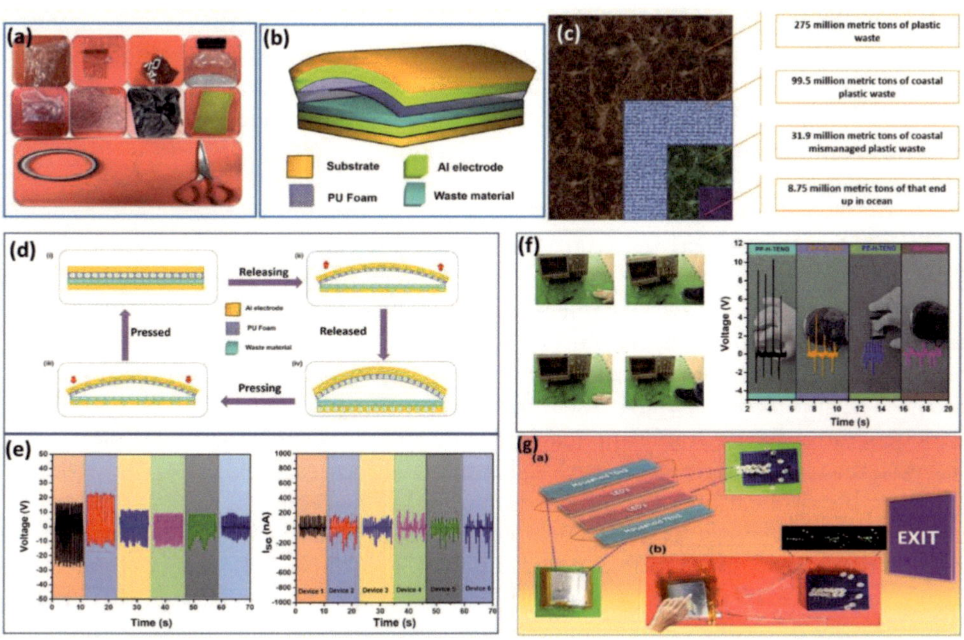

Figure 4.
(a) Fabrication materials of TENG device using recycled waste polymer wrappers from household things. (b) Schematic of the household H-TENG device. (c) Statistics of generated plastic waste all over the world. (d) Contact-separation working mechanism of H-TENG device. (e) Scavenging waste mechanical energy from H-TENG device. (f) Voltage and current response of the H-TENG devices made with different type of polymers. (g) Applications of H-TENG device showing the emergency exit with LED indication [40].

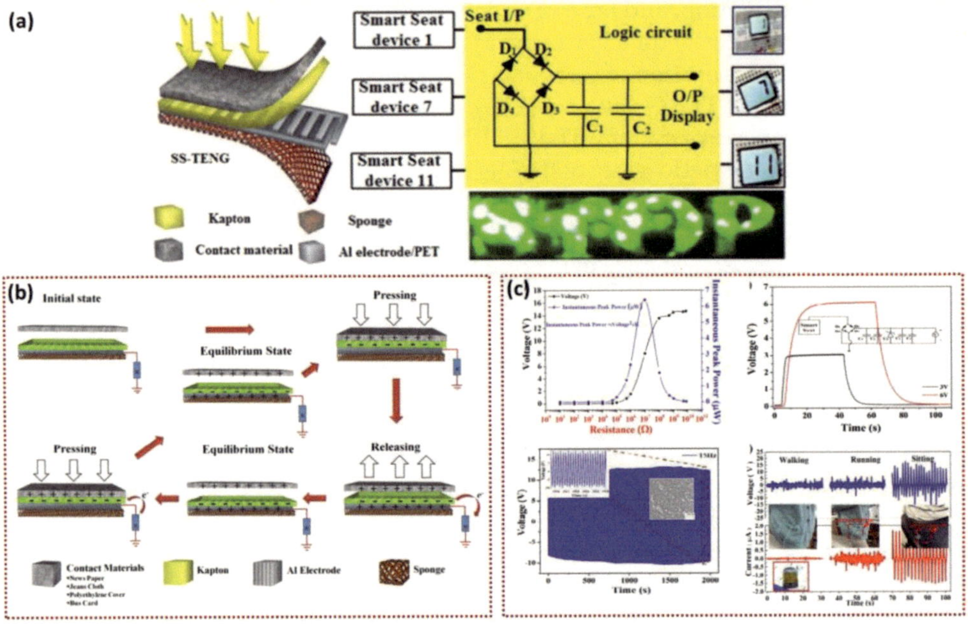

Figure 5.
(a) Schematic of SS-TENG device showing different layers and the logic circuit used for powering electronic displays and LEDs. (b) Working mechanism of SS-TENG device operating with single-electrode mode. (c) Electrical response of SS-TENG and its application as a smart seat [41].

and newspaper. The schematic of the SS-TENG device is shown in **Figure 5a**. The SS-TENG device works on a single-electrode mode triboelectrification. The working mechanism of SS-TENG device with single-electrode mode configuration is shown in **Figure 5b**. **Figure 5c** shows the electrical output response of the device showing a maximum power of 6.5 μW at 10 MΩ resistance. The device also has high stability power continuously for 2000 s without variation in the electrical response. The device is then placed in the pant pocket and measured the electrical response under walking, running, and sitting mechanical motions.

6. TENG for the portable-wearable power source

Besides scavenging biomechanical energy, TENGs can also use as a portable-wearable power sources. By making a design suitable for a portable and wearable process utilizing the appropriate working mode, TENGs can then be used for the mentioned applications. Many researchers have been working on textile-based nanogenerators for wearable applications. These devices use bio-compatible materials as triboelectric layers with the content of fabrics and making it suitable for wearable applications. The advancement in the wearable TENGs can be able to use as a wearable sensor as well as be used as a health monitoring device also. The chapter discusses a smart mobile pouch TENG (SMP-TENG) showing the portable quality of the TENG device. Along with that, a smart backpack TENG (SBP-TENG) demonstrates the energy harvesting and the wearable part.

The SMP-TENG works on the wearable part with the sliding mode. The polymeric layer made of kapton was deposited with the IDE electrode and attached in the mobile phone, which interacts with the triboelectric layer (contact materials) on the mobile pouch. The contact materials tested here are polyester, jeans, cotton, and nylon. **Figure 6a** shows the schematic of the SMP-TENG device with IDE electrode and contact materials. The working mechanism of SMP-TENG is sliding mode, with the contact materials slides over the kapton layer, as shown in **Figure 6b**. **Figure 6c** shows the electrical response of the device with contact materials. Among the four different contact materials, nylon generates a high electrical output of 150 V and 305 μA. Also, the SMP-TENG device was used to transfer the generated electrical output wirelessly using transmitter and receiver antennas.

Similarly, an SBP-TENG device also fabricated with the free-standing triboelectric made and attached to the commercial backpack. The free-standing layer here is the fabric that we wear every day. The device performance got tested with various fabric materials such as polyethylene, jeans, cloth, cotton, paper, and wool. A multi-unit SBP-TENG was constructed and attached continuously in the backpack as four units. The four devices connected in parallel electrical connection. **Figure 7a** shows the four TENG units connected in parallel, and the inset shows the backpack with TENG devices attached on the backside.

Figure 7b shows the electrical response of the device concerning the number of units connected in parallel. The voltage remains constant, and the current of the devices increases every fold with the addition of devices in parallel connection. **Figure 7c,d** shows the capacitor charging performance to increase in the TENG device under a parallel connection. The charging performance also increases, which is due to the high current with the increase in the number of devices. The electrical response for the weight of the bag is shown in **Figure 7e**. When the weight of the bag increases, the motion of the bag would be restricted; therefore, the electrical output of the device would decrease. The LED is flashed, as shown in **Figure 7f** and **g**, with the help of a capacitor and a switch. Whenever the bag moves, the capacitor gets charges due to the actuation of the TENG device; during the emergency times, the switch would be operated to give a flashlight.

Figure 7f shows the capacitor charging and discharging with the switch ON and OFF. The electrical response of the device under various contact materials is given in **Figure 7h** and **i**, where the wool as contact materials shows high output with the voltage of 250 V peak-to-peak and the current of 12 μA. **Figure 7j** and **k** shows the voltage and current response of wool material with the peak pattern. The force analysis of SBP-TENG at different weights of the backpack and their respective electrical responses are shown in **Figure 7l** and **m**. Finally, the real-time electrical response is shown in **Figure 7n** with the backpack worn by a person, and

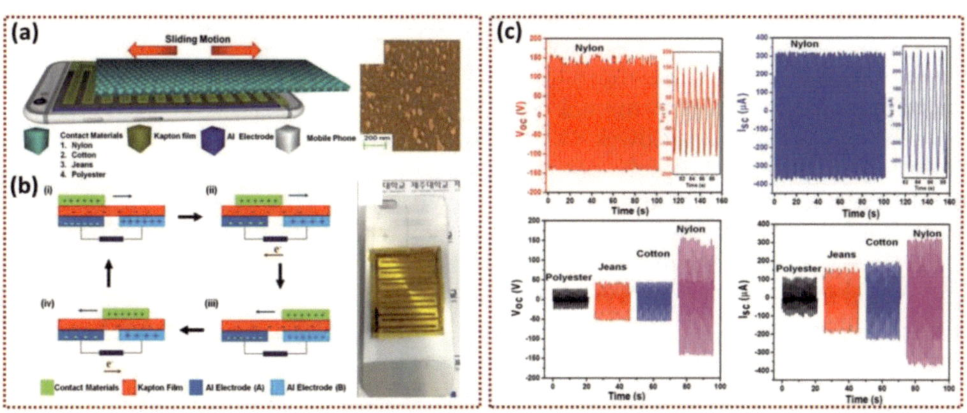

Figure 6.
(a) Schematic of a smart mobile pouch TENG device. (b) Working mechanism of SMP-TENG device operating in a free-standing triboelectrification mode. (c) Electrical response of SMP-TENG using various contact materials [42].

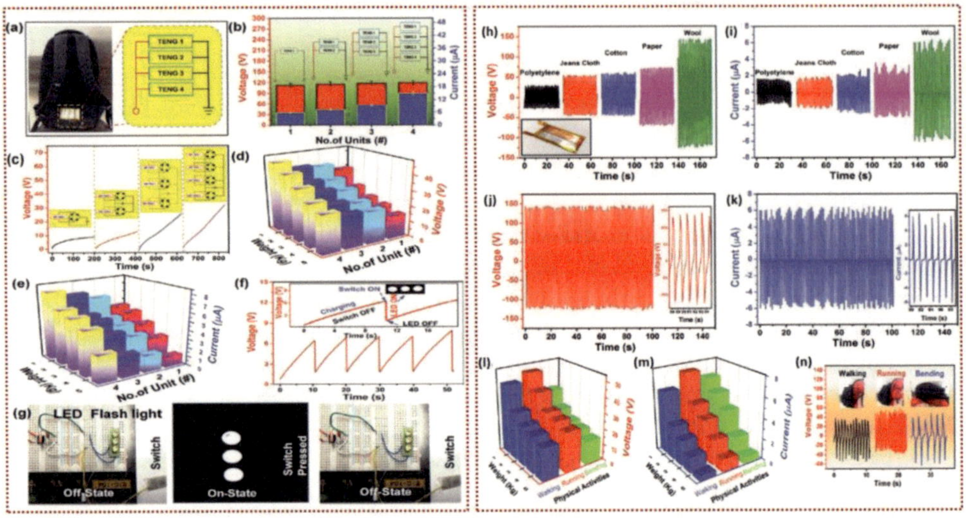

Figure 7.
(a–e) Electrical response of SBP-TENG device showing its electrical response under parallel connections with four units. (f) Cyclic stability study by charging and discharging a 1 μF capacitor. (g) Emergency LED flash using a switch and a capacitor. (h and i) Electrical response of SBP-TENG with various contact materials and the maximum response obtained with wool. (j and k) Voltage and current responses of SBP-TENG with wool. (l–n) The electrical response of SBP-TENG under biomechanical motions such as walking, running, and bending [39].

the electrical response was monitored under various biomechanical motions such as walking, running, and bending. The devices show pleasant repose under the different mechanical motions proving that the SBP-TENG is a promising candidate for scavenging biomechanical energy and for being used under emergencies.

7. Smart toy TENG

Very recently, researchers were taken a step towards the easy commercialization of TENG in a new way by making it a direct commercial product. TENGs have made as a toy which can make the children not to get addicted to electronic gadgets. The researchers have shown that the TENG could be made as a toy and sell it as a commercial product. As an idea, TENGs packed with various toys such as clapping toys, duck toys, puzzles, and computer mouse. The clapping toy makes a clapping sound, in which the contact and separation triboelectric layers were introduced between the clapping layers of the clapping toys.

The generated electric power was used to lit up green LEDs, as shown in **Figure 8a**. **Figure 8b** shows the puzzle TENG. A free-standing device made of PDMS and aluminum was attached to the puzzle board. The puzzle pieces have the other contact material, and the positions are kept different in each piece. The right piece can match with the other triboelectric layer, which placed on the exact matching area on the puzzle board. Whenever the correct piece matches, the liquid crystal display (LCD) shows "C" indicating "correct" in its display. The smart duck toy usually makes noise upon pressing its middle part. The similar concept used here by placing a TENG device placed in the middle of the duck toy connected with two LEDs in the eye portion, as shown in **Figure 8c**. The smart computer mouse uses a sliding mode triboelectrification with IDE electrodes attached to the bottom of the mouse and slides over various contact materials used in the mouse pad. This device glows a few LEDs, lit up an LCD, and charges a Li-ion battery. The improvement of

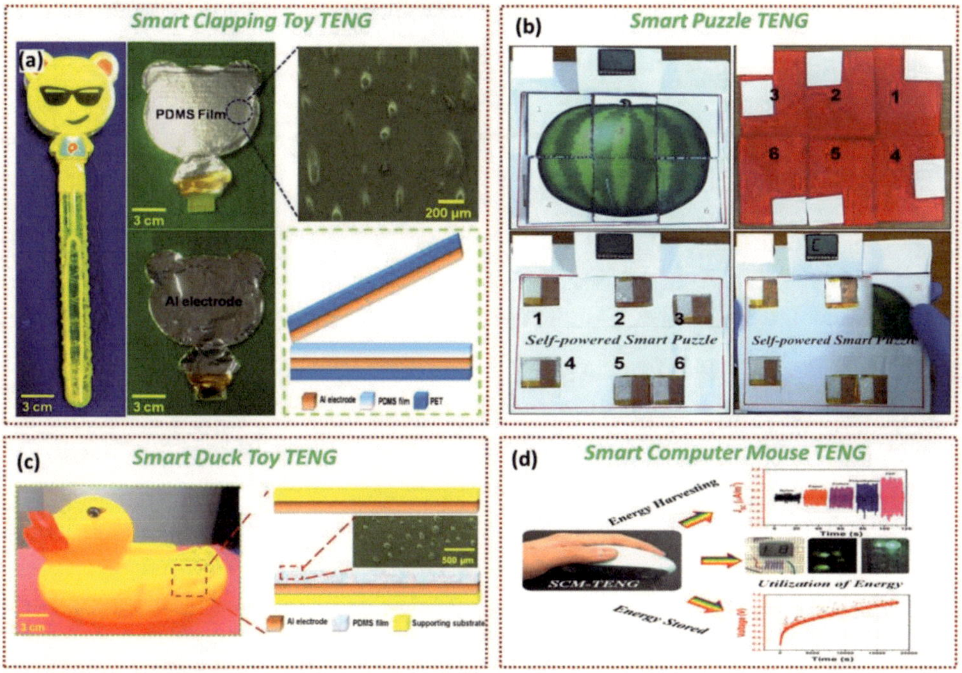

Figure 8.
(a) Schematic and device fabrication of a smart clapping toy TENG. (b) Schematic and operation of a smart puzzle TENG. (c) Schematic of a smart toy TENG. (d) Schematic and electrical response of smart mouse TENG [43, 44].

the device makes the mouse work without battery power source: the schematic and electrical response of the smart mouse TENG as shown in **Figure 8d**.

8. Edible TENG

Biodegradable and edible TENGs have recently been fabricated using edible materials. This approach utilizes the TENGs for implantable applications without being toxic to the human body and the environment. In general, TENG devices made of triboelectric materials, which are either polymer or metals. They usually create pollution in various ways depends on the toxicity levels and properties of the materials used. This section of the book shows an edible TENG made of laver (seaweed) as a triboelectric layer and an edible silver foil as an electrode. The triboelectric layer and electrode were attached to a rice sheet, which acts as a substrate for the active layer. **Figure 9a–e** shows the step-by-step fabrication process of an edible TENG device with its various layers. The TENG operated under a single-electrode mode and contacted day-to-day usage materials such as paper, tissue paper, polyvinyl chloride (PVC), and fluorinated ethylene propylene (FEP). The toxicity level and biocompatibility of the edible TENG device tested with 3-(4,5-dimethylthiazol-2-yl)-2,5-diphenyl tetrazolium bromide (MTT) assay, cell imaging, and 4′,6-diamidino-2-phenylindole (DAPI). The results prove that the triboelectric layer is biocompatible and allows the cells to grow on it. The cell growth that verifies under a confocal microscope and the imaging shows a blue color pattern in **Figure 9f**. **Figure 9g** demonstrates the real-time application of powering various electronic components such as a hygrometer, wristwatch, and LEDs using a capacitor and a switch. The electrical response and stability of the edible TENG device

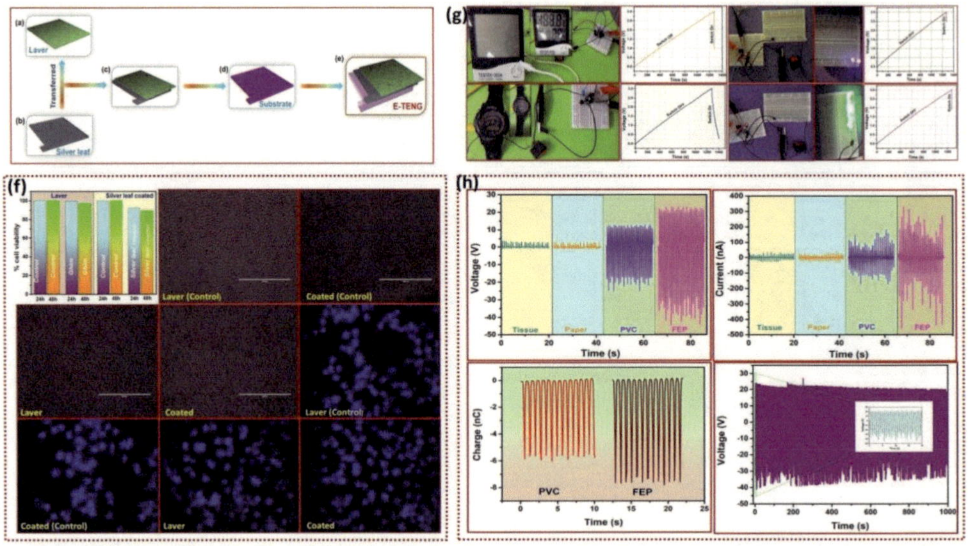

Figure 9.
(a–e) Fabrication of edible TENG device using laver as triboelectric layer and silver foil as an electrode with a rice sheet as a substrate. (f) MTT assay and cell viability test. (g) Powering up various electronic devices. (h) Electrical response analysis and stability test of the edible TENG device [45].

are shown in **Figure 9h** with the maximum output response, with its interaction with FEP. The results prove that the edible TENG is highly suitable for implantable devices and also in the applications related to biomedical electronics.

9. Fully packed-water proof TENG

There are several drawbacks of TENG devices overcome by introducing various solutions, techniques, and fabrication designs. The major drawback that TENGs are facing from the day of its invention is that its electrical performance had affected under humidity. Recently, this issue had overcome by introducing a fully packed, waterproof TENG (WP-TENG) device where it can work even underwater as well as in high humidity. **Figure 9** shows the schematic of a fully packed WP-TENG and its electrical response. The device made of nickel foam as a positive triboelectric layer, and roughness created PDMS as a negative triboelectric layer, as shown in **Figure 10a**, and the FE-SEM shows the layers in microscale in **Figure 10b–d**. The device was packed on either side with a polymer casing and sealed using a pouch laminator, as shown in **Figure 10e**. The device is extremely lightweight and weighs around 1.9 g, which is shown in **Figure 10f**. The waterproof ability of the device is tested by dipping the device underwater and is shown in **Figure 10g**. The device shows an excellent electrical response of 80 V and a maximum instantaneous power density of 4 mW/m^2. The water resistance capability of the device is shown in **Figure 10j** in which the device submerged into the water for a period of 24 h, and the corresponding response in between the intervals is measured. The response was stable for the entire time without affecting its performance. The electrical response and performance of the device are shown in **Figure 10h–j**. The electrical response to humidity is measured using a homemade humidity chamber, and the device was tested with the application of various percentages of relative humidity. The output is stable from 0% RH to 90% RH. **Figure 10k–m** shows the electrical response of humidity studies and the

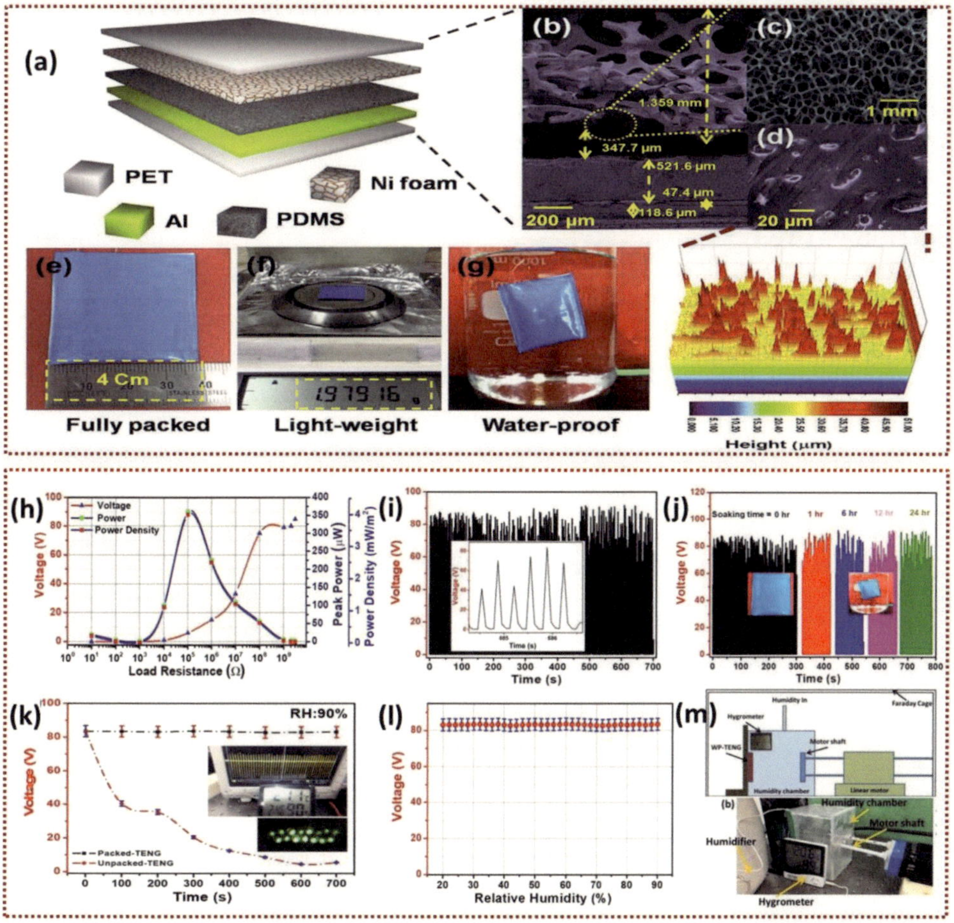

Figure 10.
(a) Schematic and layers of waterproof TENG device. (b–d) FESEM image showing the cross-section of WP-TENG device with both the layers, morphology of nickel foam and roughness PDMS. (e) Fully packed device. (f) Lightweight device weight. (g) Waterproof capability. (h) Load resistance vs. power density analysis. (i) Rectified voltage. (j) Water-resistant analysis with different time duration (k and l) humidity with output analysis. (m) Homemade humidity chamber for the analysis [34].

homemade chamber used. This study of WP-TENG shows that the device is capable of using in harsh and humid weather and environmental conditions.

10. Conclusions

In summary, the development of TENG as a sustainable, portable, and wearable power source focuses on the trend of miniaturization, energy harvesting, and self-powered electronics. The rapid advancement of energy harvesting using nanogenerators brought the possibility to power electronic devices from the waste mechanical energy in the environment. The development of TENG has reached to the bulk and scalable fabrication and started to get commercialized. Further advancement in TENG as flexible construction would be easier to harvest day-to-day mechanical energy from human motions. The construction of TENGs as a flexible power source would pave the way towards a more reliable and sustainable power source and the possibility to replace the battery shortly.

Acknowledgements

This work was supported by the Basic Science Research Program through the National Research Foundation of Korea (NRF) grant funded by the Korean Government (MSIT) (2018R1A4A1025998, 2019R1A2C3009747).

Conflict of interest

The authors declare no conflict of interest.

Author details

Venkateswaran Vivekananthan[1], Arunkumar Chandrasekhar[2], Nagamalleswara Rao Alluri[1], Yuvasree Purusothaman[1], Gaurav Khandelwal[1] and Sang-Jae Kim[1*]

1 Nanomaterials and Systems Lab, Major of Mechatronics Engineering, Faculty of Applied Energy Systems, Jeju, South Korea

2 Department of Sensors and Biomedical Engineering, Vellore Institute of Technology, India

*Address all correspondence to: kimsangj@jejunu.ac.kr

References

[1] Zhao C, Feng H, Zhang L, Li Z, Zou Y, Tan P, et al. Highly efficient In vivo cancer therapy by an implantable magnet triboelectric nanogenerator. Advanced Functional Materials. 2019;**29**(41):1808640

[2] Zheng Q, Shi B, Fan F, Wang X, Yan L, Yuan W, et al. In vivo powering of pacemaker by breathing-driven implanted triboelectric nanogenerator. Advanced Materials. 2014;**26**(33):5851-5856

[3] Moreno S, Baniasadi M, Mohammed S, Mejia I, Chen Y, Quevedo-Lopez MA, et al. Biocompatible collagen films as substrates for flexible implantable electronics. Advanced Electronic Materials. 2015;**1**(9):1500154

[4] Zheng Q, Zhang H, Shi B, Xue X, Liu Z, Jin Y, et al. In vivo self-powered wireless cardiac monitoring via implantable triboelectric nanogenerator. ACS Nano. 2016;**10**(7):6510-6518

[5] Vivekananthan V, Alluri NR, Chandrasekhar A, Purusothaman Y, Gupta A, Kim S-J. Zero-power consuming intruder identification system by enhanced piezoelectricity of $K_{0.5}Na_{0.5}NbO_3$ using substitutional doping of BTO NPs. Journal of Materials Chemistry C. 2019;7:7563-7571

[6] Park K-I, Son JH, Hwang G-T, Jeong CK, Ryu J, Koo M, et al. Highly-efficient, flexible piezoelectric PZT thin film nanogenerator on plastic substrates. Advanced Materials. 2014;**26**(16):2514-2520

[7] Fan F-R, Tian Z-Q, Lin WZ. Flexible triboelectric generator. Nano Energy. 2012;**1**(2):328-334

[8] Lee S, Hinchet R, Lee Y, Yang Y, Lin Z-H, Ardila G, et al. Ultrathin nanogenerators as self-powered/active skin sensors for tracking eye ball motion. Advanced Functional Materials. 2014;**24**(8):1163-1168

[9] Song P, Kuang S, Panwar N, Yang G, Tng DJH, Tjin SC, et al. A self-powered implantable drug-delivery system using biokinetic energy. Advanced Materials. 2017;**29**(11):1605668

[10] Vivekananthan V, Kim WJ, Alluri NR, Purusothaman Y, Abisegapriyan KS, Kim S-J. A sliding mode contact electrification based triboelectric-electromagnetic hybrid generator for small-scale biomechanical energy harvesting. Micro and Nano Systems Letters. 2019;7(1):14

[11] Wang ZL, Song J. Piezoelectric nanogenerators based on zinc oxide nanowire arrays. Science. 2006;**312**(5771):242-246

[12] Vivekananthan V, Alluri NR, Purusothaman Y, Chandrasekhar A, Kim S-J. A flexible, planar energy harvesting device for scavenging road side waste mechanical energy via the synergistic piezoelectric response of $K_{0.5}Na_{0.5}NbO_3$-$BaTiO_3$/PVDF composite films. Nanoscale. 2017;**9**(39):15122-15130

[13] Vivekananthan V, Chandrasekhar A, Alluri NR, Purusothaman Y, Joong Kim W, Kang C-N, et al. A flexible piezoelectric composite nanogenerator based on doping enhanced lead-free nanoparticles. Materials Letters. 2019;**249**:73-76

[14] Maria Joseph Raj NP, Alluri NR, Vivekananthan V, Chandrasekhar A, Khandelwal G, Kim S-J. Sustainable yarn type-piezoelectric energy harvester as an eco-friendly, cost-effective battery-free breath sensor. Applied Energy. 2018;**228**:1767-1776

[15] Alluri NR, Vivekananthan V, Chandrasekhar A, Kim S-J. Adaptable

piezoelectric hemispherical composite strips using a scalable groove technique for a self-powered muscle monitoring system. Nanoscale. 2018;**10**(3):907-913

[16] Chandrasekhar A, Vivekananthan V, Khandelwal G, Kim S-J. A Sustainable Human-Machine Interactive Triboelectric Nanogenerator towards a Smart Computer Mouse. United states: ACS Sustainable Chemistry & Engineering; 2019;**7**:7177-7182

[17] Yang Y, Guo W, Pradel KC, Zhu G, Zhou Y, Zhang Y, et al. Pyroelectric nanogenerators for harvesting thermoelectric energy. Nano Letters. 2012;**12**(6):2833-2838

[18] Bowen CR, Taylor J, LeBoulbar E, Zabek D, Chauhan A, Vaish R. Pyroelectric materials and devicesforenergyharvestingapplications. Energy & Environmental Science. 2014;7(12):3836-3856

[19] Alluri NR, Selvarajan S, Chandrasekhar A, Balasubramaniam S, Jeong JH, Kim S-J. Self powered pH sensor using piezoelectric composite worm structures derived by ionotropic gelation approach. Sensors and Actuators B: Chemical. 2016;**237**:534-544

[20] Selvarajan S, Alluri NR, Chandrasekhar A, Kim S-J. $BaTiO_3$ nanoparticles as biomaterial film for self-powered glucose sensor application. Sensors and Actuators B: Chemical. 2016;**234**:395-403

[21] Dey G, Venkateswarulu M, Vivekananthan V, Pramanik A, Krishnan V, Koner RR. Sub-picomolar recognition of Cr^{3+} through bioinspired organic–inorganic ensemble utilization. ACS Sensors. 2016;**1**(6):663-669

[22] Vivekananthan V, Alluri NR, Purusothaman Y, Chandrasekhar A, Selvarajan S, Kim S-J. Biocompatible collagen nanofibrils: An approach for sustainable energy harvesting and battery-free humidity sensor applications. ACS Applied Materials & Interfaces. 2018;**10**(22):18650-18656

[23] Chang T-H, Peng Y-W, Chen C-H, Chang T-W, Wu J-M, Hwang J-C, et al. Protein-based contact electrification and its uses for mechanical energy harvesting and humidity detecting. Nano Energy. 2016;**21**:238-246

[24] Maria Joseph Raj NP, Alluri NR, Chandrasekhar A, Khandelwal G, Kim S-J. Self-powered ferroelectric NTC thermistor based on bismuth titanate. Nano Energy. 2019;**62**:329-337

[25] Alluri NR, Chandrasekhar A, Kim S-J. Exalted electric output via piezoelectric–triboelectric coupling/sustainable butterfly wing structure type multiunit hybrid nanogenerator. ACS Sustainable Chemistry & Engineering. 2018;**6**(2):1919-1933

[26] Chandrasekhar A, Alluri NR, Saravanakumar B, Selvarajan S, Kim S-J. A microcrystalline cellulose ingrained polydimethylsiloxane triboelectric nanogenerator as a self-powered locomotion detector. Journal of Materials Chemistry C. 2017;**5**(7):1810-1815

[27] Purusothaman Y, Alluri NR, Chandrasekhar A, Vivekananthan V, Kim S-J. Direct In situ hybridized interfacial quantification to stimulate highly flexile self-powered photodetector. The Journal of Physical Chemistry C. 2018;**122**(23):12177-12184

[28] Purusothaman Y, Alluri NR, Chandrasekhar A, Vivekananthan V, Kim SJ. Regulation of charge carrier dynamics in ZnO microarchitecture-based UV/visible photodetector via photonic-strain induced effects. Small. 2018;**14**(11):1703044

[29] Wang L, Liu S, Wang Z, Zhou Y, Qin Y, Wang ZL. Piezotronic effect

enhanced photocatalysis in strained anisotropic ZnO/TiO(2) nanoplatelets via thermal stress. ACS Nano. 2016;**10**(2):2636-2643

[30] Wang S, Wang ZL, Yang Y. A one-structure-based hybridized nanogenerator for scavenging mechanical and thermal energies by triboelectric-piezoelectric-pyroelectric effects. Advanced Materials. 2016;**28**(15):2881-2887

[31] Wang H, Zhu Q, Ding Z, Li Z, Zheng H, Fu J, et al. A fully-packaged ship-shaped hybrid nanogenerator for blue energy harvesting toward seawater self-desalination and self-powered positioning. Nano Energy. 2019;**57**:616-624

[32] Vivekananthan V, Chandrasekhar A, Alluri NR, Purusothaman Y, Khandelwal G, Pandey R, et al. Fe_2O_3 magnetic particles derived triboelectric-electromagnetic hybrid generator for zero-power consuming seismic detection. Nano Energy. 2019;**64**:103926

[33] Zheng Q, Zou Y, Zhang Y, Liu Z, Shi B, Wang X, et al. Biodegradable triboelectric nanogenerator as a life-time designed implantable power source. Science Advances. 2016;**2**(3):e1501478

[34] Chandrasekhar A, Vivekananthan V, Khandelwal G, Kim SJ. A fully packed water-proof, humidity resistant triboelectric nanogenerator for transmitting Morse code. Nano Energy. 2019;**60**:850-856

[35] Purusothaman Y, Alluri NR, Chandrasekhar A, Venkateswaran V, Kim S-J. Piezophototronic gated optofluidic logic computations empowering intrinsic reconfigurable switches. Nature Communications. 2019;**10**(1):4381

[36] Shi Q, He T, Lee C. More than energy harvesting–combining triboelectric nanogenerator and flexible electronics technology for enabling novel micro−/nano-systems. Nano Energy. 2019;**57**:851-871

[37] Yang Y, Zhou YS, Zhang H, Liu Y, Lee S, Wang ZL. A single-electrode based triboelectric nanogenerator as self-powered tracking system. Advanced Materials. 2013;**25**(45):6594-6601

[38] Nie J, Wang Z, Ren Z, Li S, Chen X, Lin WZ. Power generation from the interaction of a liquid droplet and a liquid membrane. Nature Communications. 2019;**10**(1):2264

[39] Chandrasekhar A, Alluri NR, Vivekananthan V, Purusothaman Y, Kim S-J. A sustainable freestanding biomechanical energy harvesting smart backpack as a portable-wearable power source. Journal of Materials Chemistry C. 2017;**5**(6):1488-1493

[40] Khandelwal G, Chandrasekhar A, Alluri NR, Vivekananthan V, Maria Joseph Raj NP, Kim S-J. Trash to energy: A facile, robust and cheap approach for mitigating environment pollutant using household triboelectric nanogenerator. Applied Energy. 2018;**219**:338-349

[41] Chandrasekhar A, Alluri NR, Saravanakumar B, Selvarajan S, Kim S-J. Human interactive triboelectric nanogenerator as a self-powered smart seat. ACS Applied Materials & Interfaces. 2016;**8**(15):9692-9699

[42] Chandrasekhar A, Alluri NR, Sudhakaran MSP, Mok YS, Kim S-J. A smart mobile pouch as a biomechanical energy harvester towards self-powered smart wireless power transfer applications. Nanoscale. 2017;**9**(28):9818-9824

[43] Chandrasekhar A, Khandelwal G, Alluri NR, Vivekananthan V, Kim S-J. Battery-free electronic smart toys: A step toward the commercialization of sustainable triboelectric

nanogenerators. ACS Sustainable Chemistry & Engineering. 2018;**6**(5):6110-6116

[44] Chandrasekhar A, Alluri NR, Vivekananthan V, Park JH, Kim S-J. Sustainable biomechanical energy scavenger toward self-reliant kids' interactive battery-free smart puzzle. ACS Sustainable Chemistry & Engineering. 2017;**5**(8):7310-7316

[45] Khandelwal G, Minocha T, Yadav SK, Chandrasekhar A, Maria Joseph Raj NP, Gupta SC, et al. All edible materials derived biocompatible and biodegradable triboelectric nanogenerator. Nano Energy. 2019;**65**:104016

Chapter 2

Design of Electrode Materials for Stretchable Triboelectric Nanogenerators

Zhen Wen

Abstract

Triboelectric nanogenerator (TENG), a recently emerging technology that is based on the combination of triboelectric effect and electrostatic induction, has been found to be a promising strategy to harvest large amount of underutilized and low-frequency mechanical energy. One major challenge for TENGs is that the practical application requires flexible, deformable, multifunctional materials to ensure its favorable accommodation to arbitrary surfaces or moving object or harsh environment. Recent research interests mainly focus on the design and fabrication of electrode materials for TENG, making it a perfect candidate for wearable power source. In this chapter, we will introduce a couple of recent achievements regarding highly flexible/deformable TENGs based on stretchable electrodes, including geometrically designed electrode, mixture of conductive materials with elastomeric materials and intrinsically stretchable electrode, etc. In addition, we will address stretchable and self-healing electrodes of flexible TENGs for potential wearable and implantable electronics.

Keywords: triboelectric nanogenerator, electrode design, stretchable electrodes, self-healing, wearable electronics

1. Introduction

Rapid advancement of mobile and portable electronics for health care, environmental sensing and communication is a prominent trend in IoT times [1, 2]. Thus, developing corresponding wearable energy sources to satisfy the wearing demand is of great significance to enhance life quality [3]. There is abundant energy around us in various forms, such as solar energy, thermal energy, mechanical energy etc., which can be converted into electrical energy utilizing energy harvesting techniques. Among them, mechanical energy could be the most extensively distributed one. Nevertheless, mechanical energies like human motion, wind blowing, and ocean waves are almost neglected and wasted in daily life. In 2012, Fan et al. [4] demonstrated a mechanical energy harvester, triboelectric nanogenerator (TENG) based on the coupling effects of triboelectrification and electrostatic induction. Taking advantages of high voltage output, TENG can be used to power commercial digital electronics and can even directly light up LED lights.

In addition to the high electrical output performance, TENG also possesses advantages of light weight, wide-ranging material selection, simple fabrication process, etc. so that it could be regarded as a promising energy source for wearable electronics [5]. In recent years, tremendous efforts have been put into the development of flexible TENGs which can be attached on human body to harvest human motion energy like walking, patting or twisting [6–8]. Moreover, stretchability of TENG is essential since stretching behaviors happen frequently in human motions such as joint motions. So far, stretchable TENGs show extensive application in many fields, ranging from body mechanical energy harvesting, wearable sensors to monitor human health as well as biocompatible e-skin. The key challenge to fabricate stretchable TENG is to choose stretchable triboelectric materials and electrode materials which are the two main components for TENG [9]. For stretchable triboelectric materials, different kinds of elastomers (PDMS, silicone rubber, VHB, etc.) are the main source due to their softness and stretchability as well as the excellent triboelectric properties.

For stretchable electrode materials, we will introduce three main fabricating strategies in this chapter regarding different properties of conductive materials, including (i) geometrically design of rigid conductive materials, (ii) mixture of conductive materials with elastomers, (iii) intrinsically stretchable and conductive materials. In addition, we will also involve stretchable and self-healing conductive materials. Once the materials are over-stretched, self-healing capability enables the device to last longer lifetime. Meanwhile, the fabrication process of stretchable TENG based on these stretchable electrodes will be discussed as well. Herein, we only involve single-electrode TENG and contact-separation TENG. It should be noted that in some cases of contact and separation mode TENG, stretchable electrode materials can be treated as electrode materials and one of the triboelectric materials at the same time [10, 11].

2. Strategies of stretchable electrodes for TENGs

2.1 Geometrical design of rigid conductive materials

Rigid conductive materials, such as carbon papers, metal foils/wires, certain polymer thin films are all excellent candidates for the electrodes for TENG. But their low strain deformations limit their application in flexible and stretchable TENGs which is a promising developing trend for wearable electronics. In order to overcome this challenge, geometrical designs and engineering have been developed to endow these rigid conductive materials with certain stretchability. Herein, we will introduce several common geometrical engineering strategies.

Wave structural configuration can effectively accommodate large strain to avoid potential fracture of the rigid electrode materials themselves. Such wavy-shaped electrode can be achieved by deposition, coating or transferring the conductive thin film on the pre-stretched elastomeric substrate [14–17]. As **Figure 1a** shows, Wen et al. demonstrated a transparent and stretchable triboelectric nanogenerator based on wrinkled PEDOT:PSS electrode. In this work, poly(dimethylsiloxane) (PDMS) is pre-stretched and fixed on a flat glass panel, which is employed as the elastomeric substrate. Then the conductive material PEDOT:PSS film is blade-coated on this substrate. After releasing the sample from the panel, a wavy-shaped electrode is spontaneously formed (**Figure 1b**). Finally, a stretchable TENG can be obtained by inserting the Al foil as lead wire and encapsulating the whole device with additional PDMS [12]. Such a device is not only applied as body motion harvester but also as the active motion sensor attaching on human skin. Moreover, **Figure 1c** illustrates the fabrication process of a stretchable, transparent and self-healable TENG based on a buckled Ag-PEDOT electrode. The stretchability of this electrode is also achieved *via* a prestrain of substrate (H-PDMS)-transfer of conductive film

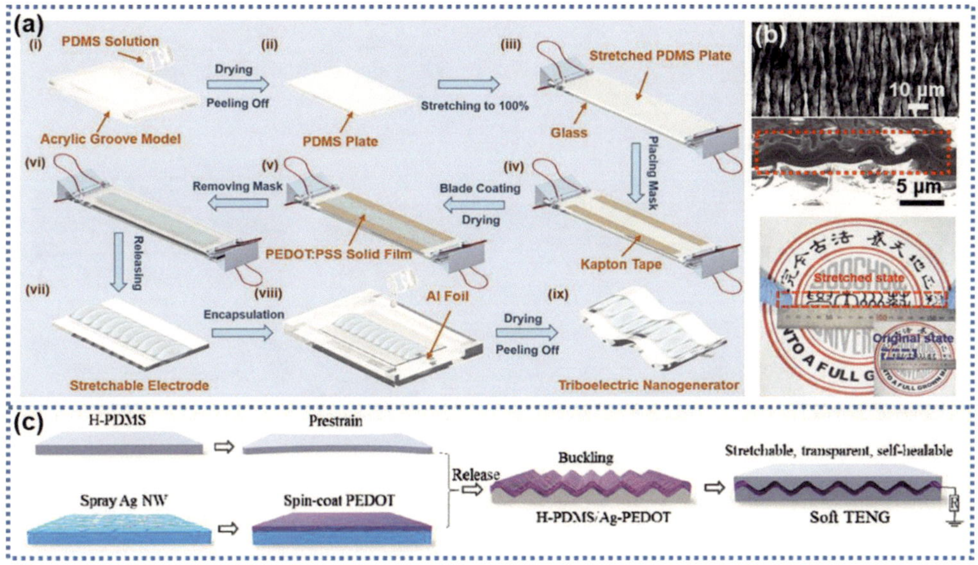

Figure 1.
Geometrical design of wave structural configuration for rigid materials. (a) Fabrication process of wavy-structured PEDOT:PSS film based triboelectric nanogenerator. (b) SEM image of a wavy-structured PEDOT:PSS film on an elastic PDMS substrate (up: Front view, down: Side view) and electrode in stretched and original state. Reproduced with permission [12]. Copyright 2018 WILEY-VCH (c) fabrication procedures of the Ag-PEDOT based stretchable electrode and the final soft TENG. Reproduced with permission [13]. Copyright 2018 American Chemical Society.

(Ag-PEDOT)-release of prestrain process. PEDOT here is spin-coated on the Ag NW film at first as an adhesive agent to anchor the networks for successful transfer [13]. Besides, in terms of self-healable ability, we will discuss about it later in this chapter which is also a significant property for stretchable electrodes. In both cases above, the ultimate stretchability of the electrodes is determined by the prestrain level of the elastomeric substrate.

Another strategy to configure stretchable electrode for TENG is to operate geometrical engineering to the existing rigid materials themselves since it is difficult for some conductive rigid materials to adhere to the prestrained substrate. As depicted in **Figure 2a** [18], the wavy structure could be realized easily through paper-folding method. The fabrication process of this stretchable TENG based on sandpaper including following steps: Firstly, the conductive paper is obtained by pencil drawing on the sandpaper. Secondly, the stretchable electrode is achieved by folding the paper into wavy structure. Lastly, the stretchable TENG is fabricated by sealing the carbon paper electrode inside silicone rubber. The ultimate stretchability of this electrode depends on the folding angle of the carbon paper. Take two folds as an example, the length of the electrode after stretching $L_{\max}$ can be describe as

$$L_{max} = {}^{L_0}/_{\sin\frac{\theta}{2}} \qquad (1)$$

Where L_0 is the original length of electrode before stretching, θ is the folding angle (**Figure 2b**). It should be noted that the ultimate stretchability of such stretchable TENG should be determined by a lower one after comparing the stretchability of silicone rubber and folded electrode. Here in Zhou's work, the folded carbon paper electrode determines the ultimate stretchability of the entire device. In addition, **Figure 2c** shows that Xie et al. [19] proposed a fiber-shaped stretchable TENG based on the geometric engineering of a steel wire electrode. The steel wire is designed into spiral structure possessing certain stretchability. Subsequently,

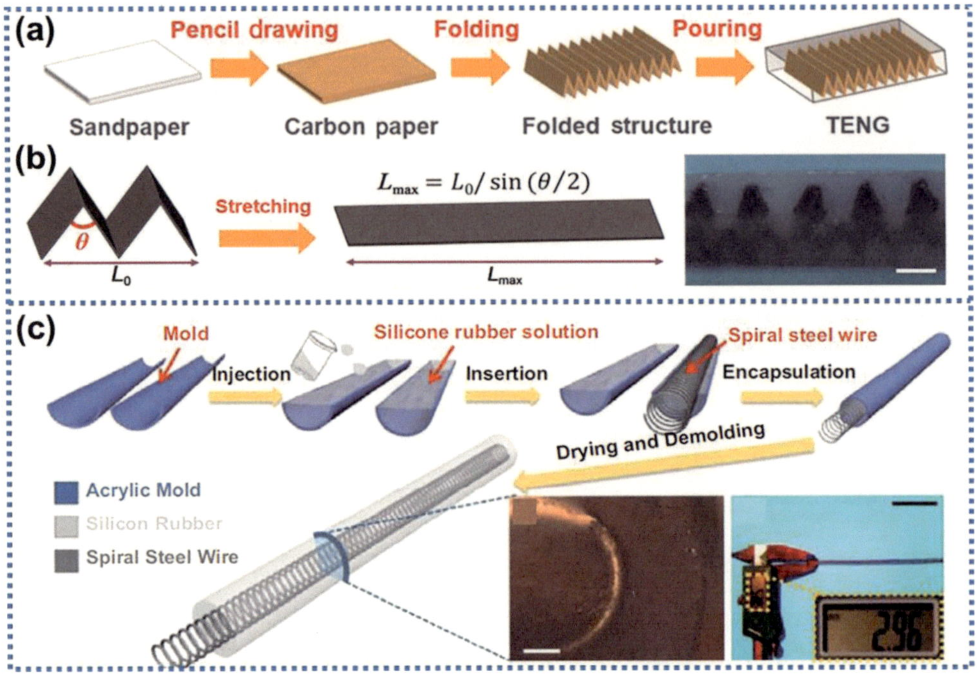

Figure 2.
Geometrical engineering of the existing rigid materials for stretchable purpose. (a) Fabrication process of wavy-structured carbon paper electrode and the final TENG. (b) the geometrical relationship between maximum stretchability and folding angle θ. reproduced with permission [18] copyright 2018 springer nature. (c) Fabrication process of the spiral steel wire electrode based TENG. Reproduced with permission [19] copyright 2019 springer.

by inserting such spiral steel wire into silicone rubber, a fiber-shaped stretchable TENG could be obtained. Similar to the previous example, the ultimate strain of this device is also determined by the strain level of the spiral steel wire since the maximum strain level of silicone rubber is much better than these geometrically designed rigid materials. Besides energy harvesting, these fiber-shaped stretchable TENGs are woven into gloves to act as gesture sensor.

The stretchability of rigid materials can also be acquired via kirigami architecture. As illustrated in **Figure 3a**, a silver-coated nylon yarn is employed as the electrode, for which the stretchability is achieved by zigzag arrangement and rhombus interlaced network. Stretchable and soft silicone rubber is selected to seal the yarn electrode network. When tensile strength is applied, the electrode network could be stretched due to gradually straightened zigzag structure and extension of rhombic region (**Figure 3b**). The ultimate stretchability of the TENG is determined by the thickness of the silicone rubber and the basic unit of network electrode including the height h and the length l (**Figure 3c**). With the h and l of 1.24 mm and d of 2.8 mm, the TENG can be stretched to 30% stain, which is eligible for e-skin and pressure sensing application [20]. Origami strategies like **Figure 3d** [21] and **Figure 3e** [22] can also endow the brittle materials with stretchability.

2.2 Mixture of conductive materials and elastomer

In this section, stretchable electrodes for TENG are acquired by directly mixing conductive materials (metal particles, carbon nanotubes, carbon black and so on) with elastomer (PDMS, silicone rubber and so on), taking advantages of the conductive property of conductive materials and the stretchability of elastomers.

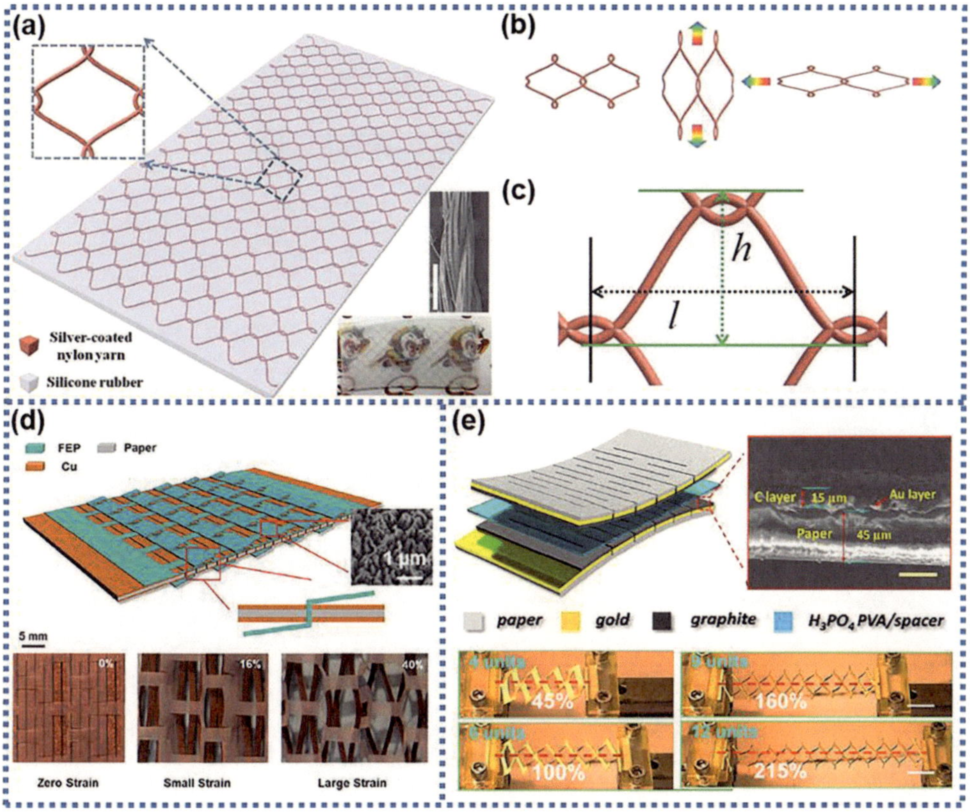

Figure 3.
Geometrical engineering of rigid materials by origami architecture. (a) Schematic illustration of the TENG with "chain-link" fence-shaped structure and rhombic unit design. (b) Schematic demonstrating the in-plane tensile behaviors of the repeated rhombic unit in the TENG system. (c) the basic cyclic unit of the yarn conducting network in the SI-TENG system. Reproduced with permission [20] copyright 2018 Wiley-VCH (d) schematic structure of paper-based TENG and the assembled device under different tensile strains. Reproduced with permission [21] copyright 2016 American Chemical Society. (e) Schematic structure of a kirigami based electrode and the related strain property. Reproduced with permission [22] copyright 2016 American Chemical Society.

Different from depositing conductive materials on prestrained elastomer surface, conductive materials in this strategy are embedded into the elastomers, making use of the fluidity of silicone rubber before solidifying.

Figure 4a illustrates a TENG based on a stretchable electrode, combining carbon black with silicone rubber. Herein, carbon black is blended with liquid silicone rubber to form a mixture. Then the mixture is coated over a piece of acrylic plate which has been processed with release agent in advance. After the mixture is solidified, a piece of stretchable conductive carbon black-silicone rubber mixture layer can be peeled off from the plate. **Figure 4b** shows the nano/microstructured surface morphologies of the stretchable electrodes in the original state. Regarding to the proposed stretchable TENG, its triboelectric materials contain one silicone rubber layer and one carbon black-silicone rubber mixture layer. Meanwhile, the stretchable mixture layers act as the electrodes for TENG. An air gap is created between the two layers to realize contact and separation process. With the increasing strain of TENG, electrical output is enhanced resulting from the increasing displacement of two triboelectric layers (**Figure 4c**) [23]. **Figure 4d** demonstrates another super-stretchable TENG via percolating networks effect of silver nanowires and silicone rubber. The fabrication process is as follows: Firstly, Ag NWs suspension is drop-coated on an acrylic plate with a specific shape molded by kapton tapes. Next, the

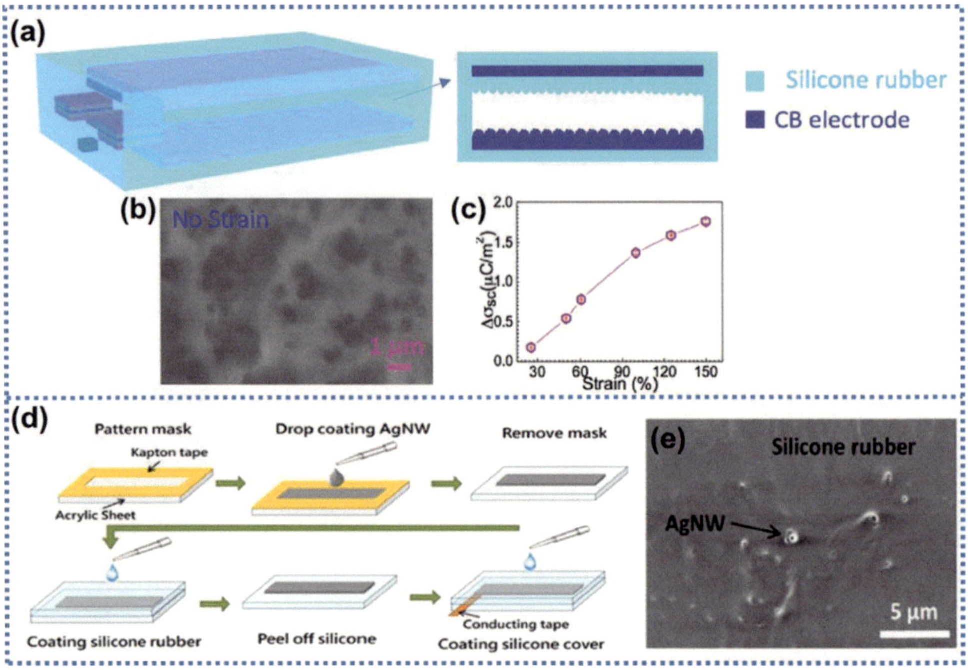

Figure 4.
Mixture of conductive materials and elastomer. (a) Schematic diagram showing the detailed structure of the stretchable TENG. (b) SEM images of the surface morphology of the TENG's electrode in the original state. (c) Relationship between the $\Delta\sigma_{sc}$ and the elongation of the device. Reproduced with permission [23] copyright 2016 American Chemical Society (d) fabrication process of the silver nanowire- silicone rubber electrode based TENG. (e) SEM image of stretchable electrode. Reproduced with permission [24] copyright 2016 Wiley-VCH.

tapes are removed from the sheet and liquid silicone rubber is cast onto the as-fabricated Ag NW network. After solidifying of silicone rubber, Ag NWs-silicone rubber film is peeled off from the acrylic plate. Finally, stretchable TENG is obtained by curing another liquid silicone rubber again over the film with a copper tape as conductive lead for assembling the Ag NW-silicone rubber layer. It is clearly shown in **Figure 4e** that Ag NWs are embedded in the silicone rubber because interspace between Ag NWs allows silicone rubber to percolate into the network structure. The proposed Ag NWs-silicone rubber mixture layer exhibits outstanding stretchability, so as for the fabricated TENG, reaching up to 300% strain [24].

In the work of Lim et al., they fabricated a stretchable TENG based on highly stretchable Au NS-embedded PDMS electrode. As shown in **Figure 5a**, Au NS multilayer film was prepared by transferring monolayer of the Au NSs onto a Teflon coated Si wafer for several times. After each transfer, thermal annealing is conducted to increase the contact between stacked Au NSs. Then, liquid PDMS is poured to the as-prepared Au NS film, percolating into the interspace between the NSs. After curing, the Au NS-embedded PDMS film can be peeled off from the Si wafer. In this stretchable electrode film, Au NSs are densely entangled and partially embedded in the PDMS matrix (**Figure 5b**). Different from directly depositing conductive materials on the surface of elastomer, embedding conductive materials into elastomers presents a more stable and slight resistance change for stretchable electrodes especially under larger strain (>30%) (**Figure 5c–e**). Such TENG is employed to detect movement of fingers [25]. Li et al. proposed electrospinning technique as an effective strategy for the fabrication of conductive materials-elastomer mixed stretchable electrode.

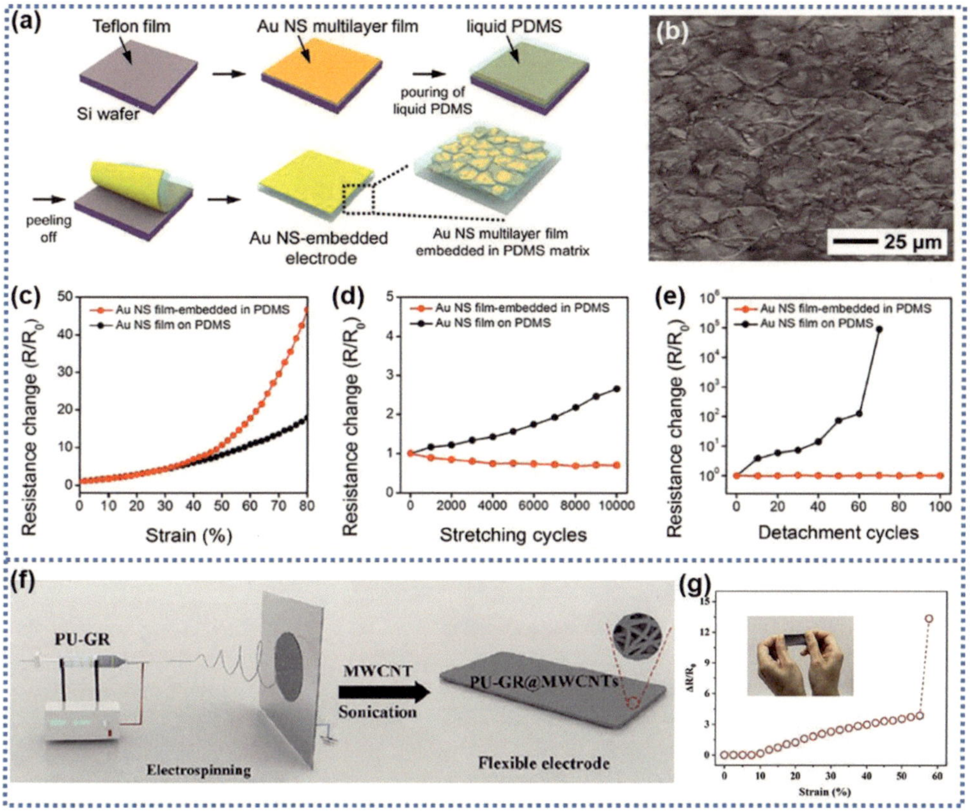

Figure 5.
Mixture of conductive materials and elastomer. (a) Fabrication process of the Au NS-embedded electrode. (b) Top-view FE-SEM image of the Au NS-embedded electrode. (c) Change in resistance for Au NS electrodes in response to tensile strain. (d) Variations of R/Ro for Au NS electrodes during 10,000 cycles of stretching at a 30% strain. (e) Variations of R/Ro for Au NS electrodes during 100 cycles of the scotch-tape detachment test. Reproduced with permission [25] copyright 2017 Elsevier (f) fabrication process of PU-GR@MWCNTs stretchable electrode. (g) Resistance change of the electrode under different strains. Reproduced with permission [26] copyright 2019 Elsevier.

As **Figure 5f** shows, polyurethane (PU) and graphene (GR) are mixed uniformly as the solution for electrospinning. Then, the prepared PU-GR film is sonicated in the solution of MWCNTs to reduce its internal resistance. Consequently, MWCNTs coated PU-GR (PU-GR@MWCNTs) stretchable electrode is obtained. Such stretchable electrode can withstand a strain of about 15% without a remarkable resistance increase (**Figure 5g**). Researchers also attached this TENG on the leaves to monitor wind speed in the natural environment [26]. In contrast with previously described wavy-structure electrodes whose stretchability mainly depends on the pre-strain level, electrodes based on percolating networks or mixture of conductive materials with elastomers display strong dependence on the capacity of the overlapping of conductive materials in response to the applied strain.

2.3 Intrinsically stretchable conductive materials

Considering the above two strategies, although stretchability can be acquired for electrodes, it is greatly limited by the geometrical engineering technique and conductive pathways of conductive materials when embedded into elastomers.

In this section, we will introduce intrinsically stretchable conductive materials including liquid-state materials with extremely low Young's modulus and prepared

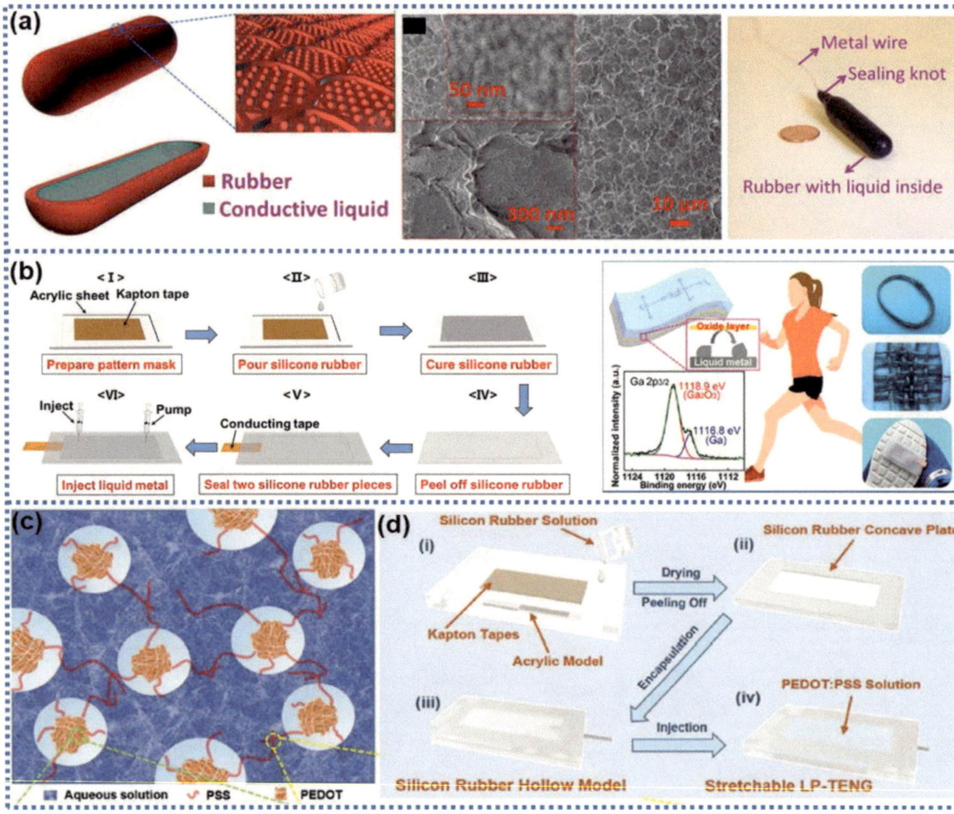

Figure 6.
Conductive liquid-state materials based stretchable electrode. (a) The structure of conductive liquid based TENG unit. Reproduced with permission from the authors [27] (b) fabrication process of liquid metal based TENG. Reproduced with permission [28] copyright 2018 American Chemical Society (c) the PEDOT:PSS particles dispersed in the aqueous solution. (d) Fabrication process of liquid PEDOT:PSS based TENG. Reproduced with permission [29] copyright 2019 the Royal Society of Chemistry.

conductive gels. As shown in **Figure 6a**, a stretchable TENG is composed of conductive liquid inside and a rubber layer outside with etched nanostructures. In the fabrication process, conductive liquid is injected into a hollow cylindrical rubber layer with one terminal open and the other one sealed. After injection, a copper wire is inserted into the liquid and a knot is formed to seal the open terminal. NaCl is used here as the conductive liquid electrode which has certain conductivity and infinite stretchability due to its fluidity at room temperature [27]. Ultimate stretchability only relies on the stretchability of the rubber, which means liquid-state conductive materials can exercise the maximum potential for stretchable electrodes.

However, the performance of TENG is limited due to the low conductivity of ion solution. Thus, in Yang et al.'s work, [28] they demonstrate a super-stretchable and high-performance TENG using liquid metal (Galinstan) electrode which exhibit low Young's modulus, high conductivity at room temperature. Silicone rubber is chosen as the triboelectric layer and encapsulating layer. **Figure 6b** shows the fabrication process of this stretchable bulk-shaped TENG: A silicone rubber model is prepared by encapsulating two concave silicone rubber plates together with a conducting tape inserted to form a cavity in the middle. Single-mode stretchable TENG is obtained after Galinstan is injected into the cavity. The as-prepared stretchable TENG has an ultimate stretchability of 300% depending on the silicone rubber. Meanwhile, its electrical output performance is much higher than the ion-solution based stretchable TENG.

Similar to this work, Shi et al. [29] utilize the liquid-state PEDOT:PSS as electrode for fabricating TENG with silicone rubber as triboelectric layer as well as encapsulating layer. Conductive polymer, PEDOT:PSS are commonly used in organic solar cells, organic light-emitting diodes (OLED) due to its transparency and conductivity while in solid state. As illustrated in **Figure 6c**, PEDOT and PSS particles are dispersed in the aqueous solution and combined by electrostatic attraction. Significant conductivity and stretchability of PEDOT:PSS electrode is attributed to the conjugated bond of PEDOT like other conductive polymers. **Figure 6d** shows similar fabrication process of this polymer liquid electrode based TENG as liquid metal based TENG. Such prepared liquid PEDOT:PSS TENG can also be stretched to 300% at most without any crack.

Moreover, hydrogels, which contains hydrophilic polymer networks swollen with water or ionic aqueous solution, are intrinsically stretchable materials, so that they can be employed as stretchable conductive electrodes for TENG. For example, stretchable electrodes based on polyacrylamide (PAAm)-(lithium chloride) LiCl hydrogel is proposed by Pu et al. for application in the stretchable TENG [30]. PAAm powder is added into LiCl solution, followed by adding N, N′-methylenebisacrylamide, ammonium persulfate, and N, N, N′, N′-tetramethylethylenediamine orderly into the solution. Then the solution is transferred into a mold. After treated at 50°C for 2 hours, conductive and stretchable hydrogel is formed. Using elastomer VHB/PDMS as triboelectric materials as well as encapsulating materials, the hydrogel is sealed inside the elastomer to form single-electrode mode TENG (**Figure 7a**). Such VHB-hydrogel based TENG can be stretched to 1000% of original length while the PAAm-LiCl itself can be stretched to 1400% (**Figure 7b**). It means that the conductive hydrogel electrode present enough stretchability for the fabrication of stretchable TENG. The stretchability is only controlled by the encapsulation materials. Zhao et al. [31] utilize PAMPS ionogel as the electrode for stretchable TENG, where the conductivity of the ionogel is

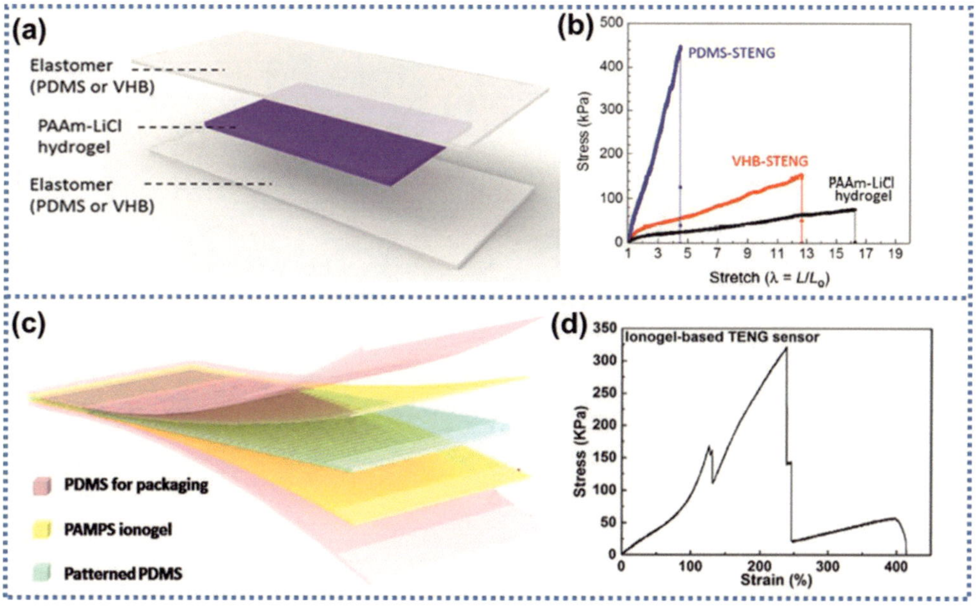

Figure 7.
Conductive gels based stretchable electrode. (a) The structure of PAAm-LiCl hydrogel electrode based TENG. (b) Tensile test of the PAAm-LiCl hydrogel, PDMS-STENG and VHB-TENG. Reproduced with permission of the authors [30] (c) the structure of PAMPS ionogel electrode based TENG. (d) Tensile test of the ionogel-based TENG. Reproduced with permission [31] copyright 2019 Elsevier.

attributed to the nonvolatility of ILs. As shown in **Figure 7c**, TENG works in contact and separation mode with patterned PDMS acting as one triboelectric material and ionogel acting as the other triboelectric material. Distance between two triboelectric materials is created due to triangular stripes on the patterned PDMS. The three layers (one patterned PDMS and two ionogel layers) are sealed inside two smooth PDMS films. Such device has an ultimate stretchability of about 125% which is determined by the ionogel film which has lower strain level compared with the patterned triboelectric PDMS and smooth packaging PDMS (**Figure 7d**). Due to the transparency, stretchability and pressure sensitive property of TENG, touching and pressure sensing are demonstrated in both work in **Figure 7**.

2.4 Self-healing conductive materials

When the electrodes are over stretched, fractures would occur to influence its conductivity as well as the output performance of TENG. Hence, self-healing ability is required for electrode materials so that fractured structures can be repaired to maintain performance and elongate the lifetime of devices The self-healing capability of electrode can be attributed several mechanisms, such as the containing of self-healing agents inside microcapsules, the use of dynamic bonds.

As **Figure 8a** shows, Parida et al. [32] prepared a stretchable and self-healing TENG based on slime ionic electrode. VHB adhesive tape, acting as the substrate and encapsulation material also possesses stretchability and self-healing capability. Although silicone rubber is not intrinsically self-healable, the self-healing process of

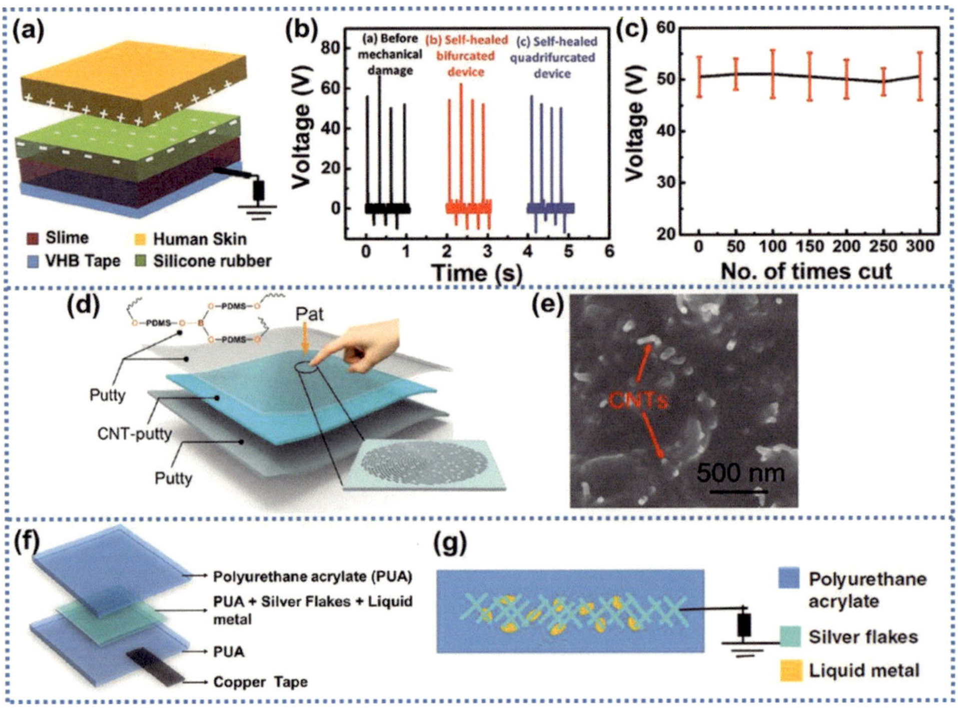

Figure 8.
Self-healing conductive stretchable materials based electrode. (a) The structure of slime electrode based self-healing TENG. (b) Voltage output of the self-healing TENG before damaging and after healing. (c) Stability of the TENG after multiple cut. Reproduced with permission [32] copyright 2017 Wiley-VCH (e) SEM image of CNTs embedded putty. Reproduced with permission [33] copyright 2019 American Chemical Society (f) structure of PUA-silver flake-liquid metal electrode based TENG. (g) Schematic diagram of the stretchable and healable triboelectric nanogenerator indicating silver flakes and liquid metal particles are embedded in the PUA matrix. Reproduced with permission from the authors [34].

slime electrode and VHB tape can bring the damaged silicone rubber film back into contact. Meanwhile, it would not influence triboelectrification with skin or electrostatic induction in electrode because charges over silicone rubber are immovable. Thus, the entire device can be regarded as self-healing TENG. Ultimate stretchability of the device can be as high as 700%. The damaged device can be successfully healed at room temperature without any external stimuli. **Figures 8b** and **c** show that output performance of TENG before damage and after self-healing is comparable to each other, even damaged and recovered for 300 times.

Chen et al. [33] demonstrate a stretchable and self-healing TENG utilizing viscoelastic supramolecular polymer (Putty) as triboelectric materials, mixture of carbon nanotubes and Putty as electrode materials (**Figure 8d**). The percolating method of configuring stretchable electrode is introduced in previous part. **Figure 8e** shows the morphology of the CNT embedded putty. Self-healing capability of CNT-Putty electrode is acquired because of reversible dynamic hydrogen bonds and dative bonds between boron and oxygen in putty, enabling the stretchable TENG to recover from damage at room temperature in 3 min.

Another example of self-healing and stretchable electrode for TENG is fabricated by mixing polyurethane acrylate (PUA) solution, liquid metal particles and silver flakes together. And then the mixture is drop casted and sandwiched between two PUA films. A self-healing and stretchable TENG can be developed after UV cures (**Figure 8f**). This TENG shows the highest stretchability (2500%), compared to previously reported TENGs. Liquid metal here provides effective anchoring between silver flakes in the PUA matrix, ensuring stable electrical performance (**Figure 8g**). Both stretchability and self-healing capability of the device is also attributed to the supramolecular hydrogen-bonding of PUA. It takes about 24 h for this TENG to recover at room temperature [34].

3. Conclusions

Overall, we have introduced three functional strategies in this chapter to fabricate the stretchable electrodes for flexible and wearable TENGs. In the first strategy, we present several methods considering different material properties, including prestrain-coating-release method for conductive materials that can be coated and cured on the surface of stretchable elastomer, structure design and origami architecture method for those existing conductive materials. In the second strategy, conductive and stretchable electrodes are fabricated by embedding conductive materials into the elastomer. In the third strategy, intrinsically stretchable electrode like liquid-state electrode and conductive gel are presented. Besides, we also discussed about the self-healing and stretchable electrodes for TENG application. These stretchable electrodes ensure the great flexibility of TENG so that it can be further applied in wearable electronics.

Acknowledgements

The work was supported by National Natural Science Foundation of China (No. 61804103) and Natural Science Foundation of Jiangsu Province of China (Nos. BK20170343). This work is also supported by Collaborative Innovation Center of Suzhou Nano Science & Technology, the Priority Academic Program Development of Jiangsu Higher Education Institutions (PAPD), the 111 Project and Joint International Research Laboratory of Carbon-Based Functional Materials and Devices.

Conflict of interest

The authors declare no conflict of interest.

Author details

Zhen Wen
Institute of Functional Nano and Soft Materials (FUNSOM), Soochow University, Suzhou, China

Address all correspondence to: wenzhen2011@suda.edu.cn

References

[1] Cheng T, Zhang Y, Lai WY, Huang W. Stretchable thin-film electrodes for flexible electronics with high deformability and stretchability. Advanced Materials. 2015;**27**(22):3349-3376. DOI: 10.1002/adma.201405864

[2] Yang JC, Mun J, Kwon SY, Park S, Bao Z, Park S. Electronic skin: Recent progress and future prospects for skin-attachable devices for health monitoring, robotics, and prosthetics. Advanced Materials. 2019;**31**:e1904765. DOI: 10.1002/adma.201904765

[3] Fan FR, Tang W, Wang ZL. Flexible nanogenerators for energy harvesting and self-powered electronics. Advanced Materials. 2016;**28**(22):4283-4305. DOI: 10.1002/adma.201504299

[4] Fan FR, Tian ZQ, Wang ZL. Flexible triboelectric generator. Nano Energy. 2012;**1**(2):328-334. DOI: 10.1016/j.nanoen.2012.01.004

[5] Sun N, Wen Z, Zhao FP, Yang YQ, Shao HY, Zhou CJ, et al. All flexible electrospun papers based self-charging power system. Nano Energy. 2017;**38**:210-217. DOI: 10.1016/j.nanoen.2017.05.048

[6] Xiong J, Cui P, Chen X, Wang J, Parida K, Lin MF, et al. Skin-touch-actuated textile-based triboelectric nanogenerator with black phosphorus for durable biomechanical energy harvesting. Nature Communications. 2018;**9**(1):4280. DOI: 10.1038/s41467-018-06759-0

[7] Wang L, Daoud WA. Highly flexible and transparent polyionic-skin triboelectric nanogenerator for biomechanical motion harvesting. Advanced Energy Materials. 2018;**9**:1803183. DOI: 10.1002/aenm.201803183

[8] Wang X, Yin Y, Yi F, Dai K, Niu S, Han Y, et al. Bioinspired stretchable triboelectric nanogenerator as energy-harvesting skin for self-powered electronics. Nano Energy. 2017;**39**:429-436. DOI: 10.1016/j.nanoen.2017.07.022

[9] Ryu H, Kim S-W. Recent development of the triboelectric properties of the polymer: A review. Advanced Materials Letters. 2018;**9**(7):462-470. DOI: 10.5185/amlett.2018.1869

[10] Chen J, Guo H, He X, Liu G, Xi Y, Shi H, et al. Enhancing performance of triboelectric nanogenerator by filling high dielectric nanoparticles into sponge pdms film. ACS Applied Materials & Interfaces. 2016;**8**(1):736-744. DOI: 10.1021/acsami.5b09907

[11] Wang J, Wen Z, Zi YL, Zhou PF, Lin J, Guo HY, et al. All-plastic-materials based self-charging power system composed of triboelectric nanogenerators and supercapacitors. Advanced Functional Materials. 2016;**26**(7):1070-1076. DOI: 10.1002/adfm.201504675

[12] Wen Z, Yang YQ, Sun N, Li GF, Liu YN, Chen C, et al. A wrinkled pedot:Pss film based stretchable and transparent triboelectric nanogenerator for wearable energy harvesters and active motion sensors. Advanced Functional Materials. 2018;**28**(37):1803684. DOI: 10.1002/adfm.201803684

[13] Sun J, Pu X, Liu M, Yu A, Du C, Zhai J, et al. Self-healable, stretchable, transparent triboelectric nanogenerators as soft power sources. ACS Nano. 2018;**12**(6):6147-6155. DOI: 10.1021/acsnano.8b02479

[14] Ho X, Cheng CK, Tey JN, Wei J. Biaxially stretchable transparent conductors that use nanowire networks. Journal of Materials Research. 2014;**29**(24):2965-2972. DOI: 10.1557/jmr.2014.338

[15] Pyo JB, Kim BS, Park H, Kim TA, Koo CM, Lee J, et al. Floating compression of ag nanowire networks for effective strain release of stretchable transparent electrodes. Nanoscale. 2015;7(39):16434-16441. DOI: 10.1039/c5nr03814f

[16] Jin Y, Hwang S, Ha H, Park H, Kang S-W, Hyun S, et al. Buckled au@pvp nanofiber networks for highly transparent and stretchable conductors. Advanced Electronic Materials. 2016;**2**(2):1500302. DOI: 10.1002/aelm.201500302

[17] Xu F, Wang X, Zhu Y, Zhu Y. Wavy ribbons of carbon nanotubes for stretchable conductors. Advanced Functional Materials. 2012;**22**(6):1279-1283. DOI: 10.1002/adfm.201102032

[18] Zhou CJ, Yang YQ, Sun N, Wen Z, Cheng P, Xie XK, et al. Flexible self-charging power units for portable electronics based on folded carbon paper. Nano Research. 2018;**11**(8):4313-4322. DOI: 10.1007/s12274-018-2018-8

[19] Xie L, Chen X, Wen Z, Yang Y, Shi J, Chen C, et al. Spiral steel wire based fiber-shaped stretchable and tailorable triboelectric nanogenerator for wearable power source and active gesture sensor. Nano-Micro Letters. 2019;**11**(1):39. DOI: 10.1007/s40820-019-0271-3

[20] Dong K, Wu Z, Deng J, Wang AC, Zou H, Chen C, et al. A stretchable yarn embedded triboelectric nanogenerator as electronic skin for biomechanical energy harvesting and multifunctional pressure sensing. Advanced Materials. 2018;**30**:43, e1804944. DOI: 10.1002/adma.201804944

[21] Wu C, Wang X, Lin L, Guo H, Wang ZL. Paper-based triboelectric nanogenerators made of stretchable interlocking kirigami patterns. ACS Nano. 2016;**10**(4):4652-4659. DOI: 10.1021/acsnano.6b00949

[22] Guo H, Yeh MH, Lai YC, Zi Y, Wu C, Wen Z, et al. All-in-one shape-adaptive self-charging power package for wearable electronics. ACS Nano. 2016;**10**(11):10580-10588. DOI: 10.1021/acsnano.6b06621

[23] Yi F, Wang J, Wang X, Niu S, Li S, Liao Q, et al. Stretchable and waterproof self-charging power system for harvesting energy from diverse deformation and powering wearable electronics. ACS Nano. 2016;**10**(7):6519-6525. DOI: 10.1021/acsnano.6b03007

[24] Lai Y-C, Deng J, Niu S, Peng W, Wu C, Liu R, et al. Electric eel-skin-inspired mechanically durable and super-stretchable nanogenerator for deformable power source and fully autonomous conformable electronic-skin applications. Advanced Materials. 2016;**28**(45):10024-10032. DOI: 10.1002/adma.201603527

[25] Lim G-H, Kwak SS, Kwon N, Kim T, Kim H, Kim SM, et al. Fully stretchable and highly durable triboelectric nanogenerators based on gold-nanosheet electrodes for self-powered human-motion detection. Nano Energy. 2017;**42**:300-306. DOI: 10.1016/j.nanoen.2017.11.001

[26] Li X, Jiang C, Zhao F, Lan L, Yao Y, Yu Y, et al. Fully stretchable triboelectric nanogenerator for energy harvesting and self-powered sensing. Nano Energy. 2019;**61**:78-85. DOI: 10.1016/j.nanoen.2019.04.025

[27] Yi F, Wang X, Niu S, Li S, Yin Y, Dai K, et al. A highly shape-adaptive, stretchable design based on conductive liquid for energy harvesting and self-powered biomechanical monitoring. Science Advances. 2016;**2**(6):e1501624. DOI: 10.1126/sciadv.1501624

[28] Yang Y, Sun N, Wen Z, Cheng P, Zheng H, Shao H, et al. Liquid-metal-based super-stretchable and structure-designable triboelectric nanogenerator

for wearable electronics. ACS Nano. 2018;**12**(2):2027-2034. DOI: 10.1021/acsnano.8b00147

[29] Shi J, Chen X, Li G, Sun N, Jiang H, Bao D, et al. A liquid pedot:Pss electrode-based stretchable triboelectric nanogenerator for a portable self-charging power source. Nanoscale. 2019;**11**(15):7513-7519. DOI: 10.1039/c9nr01271k

[30] Pu X, Liu M, Chen X, Sun J, Du C, Zhang Y, et al. Ultrastretchable, transparent triboelectric nanogenerator as electronic skin for biomechanical energy harvesting and tactile sensing. Science Advances. 2017;**3**(5):e1700015. DOI: 10.1126/sciadv.1700015

[31] Zhao G, Zhang Y, Shi N, Liu Z, Zhang X, Wu M, et al. Transparent and stretchable triboelectric nanogenerator for self-powered tactile sensing. Nano Energy. 2019;**59**:302-310. DOI: 10.1016/j.nanoen.2019.02.054

[32] Parida K, Kumar V, Jiangxin W, Bhavanasi V, Bendi R, Lee PS. Highly transparent, stretchable, and self-healing ionic-skin triboelectric nanogenerators for energy harvesting and touch applications. Advanced Materials. 2017;**29**(37):1702181. DOI: 10.1002/adma.201702181

[33] Chen Y, Pu X, Liu M, Kuang S, Zhang P, Hua Q, et al. Shape-adaptive, self-healable triboelectric nanogenerator with enhanced performances by soft solid-solid contact electrification. ACS Nano. 2019;**13**(8):8936-8945. DOI: 10.1021/acsnano.9b02690

[34] Parida K, Thangavel G, Cai G, Zhou X, Park S, Xiong J, et al. Extremely stretchable and self-healing conductor based on thermoplastic elastomer for all-three-dimensional printed triboelectric nanogenerator. Nature Communications. 2019;**10**(1):2158. DOI: 10.1038/s41467-019-10061-y

Section 2

Piezoelectric Nanogenerator

Chapter 3

Development of Vibration Piezoelectric Harvesters by the Optimum Design of Cantilever Structures

Prateek Asthana and Gargi Khanna

Abstract

Piezoelectric energy harvesting is a way of converting waste mechanical energy into usable electrical form. The selection of mechanical devices for conversion of mechanical to electrical energy is a significant part of vibration energy harvesting. The articles provide designing and optimization of a cantilever piezoelectric energy harvester. At first, is the selection of best mechanical device for energy harvesting application. A cantilever without proof mass is then analyzed for the selection of substrate, and piezoelectric material also plays a key role in the performance of the device. Aluminum is selected as a substrate, while zinc oxide acts as the piezoelectric layer. Addition of proof mass reduces the resonant frequency of the device to about 51 Hz as compared to 900 Hz for an aluminum cantilever beam. An electromechanical study shows an active conversion of mechanical input energy to electrical output energy. Power frequency response functions of the resultant structure are able to generate 0.47 mW power having 6.8 μA current at 1 g input acceleration.

Keywords: piezoelectric, cantilever, energy harvesting, microscale, vibration-based energy harvesting

1. Introduction

The electronic world is moving toward smarter, smaller, portable, and reliable devices. With the range of evolution in microelectronics and wireless technologies, wireless sensor network is one of the most growing areas of research. Wireless sensor network has a wide range of application in defense and military to healthcare and industrial and structural health and environmental monitoring. WSN consists of low-power microsensors, which collect data from the environment and transmit to the base controlling station [1]. These nodes are spread over a large area forming an integrated network. In military purposes for monitoring of remote area and unreachable terrains, sensor networks are used. Environmental monitoring for temperature and CO_2 emission can be analyzed using sensor networks. These sensor networks find a wide range of applications in healthcare sector, finding applications in wearable smart clothing and in monitoring patients' blood pressure. Monitoring the health and stability of civil structures like bridges and buildings are few other areas of applications for sensor nodes. These nodes have three important components, i.e., transmitter, sensing device, and power supply. The transmitter is utilized in communicating with controlling node, while the sensing device senses environmental characteristics [2]. A sensor node needs an uninterrupted power supply to operate throughout its lifetime, and hence the life of a sensor node is governed by the battery. These energy harvesters could provide power to sensor nodes for transmitting and monitoring purposes. Battery maintenance and refurbishment are not possible in WSN. Hence, energy harvesting to power these sensor nodes is the most desirable area of research. Microscale energy harvesting (micro-nano watt) can be used to drive small microscale nodes for environmental monitoring and medical health monitoring purposes. The focus of micro/nano-scale energy harvesting is using energy from the environment [3]. Energy from the environment, harvested in the form of vibrations, is generally small, as is the power requirement for microscale devices; hence these ambient vibrations due to their ubiquitous existence everywhere are a popular source of energy harvesting [4]. There has been a great focus in the past few years in the design and development of vibration-based micro-energy generation to power low-power wireless devices. Ambient vibration provides a viable option for harvesting energy through piezoelectric mode [5].

2. Device design

Device geometry plays an important role in determining the output power of the device. The amount of power produced by the harvester depends on the amount of

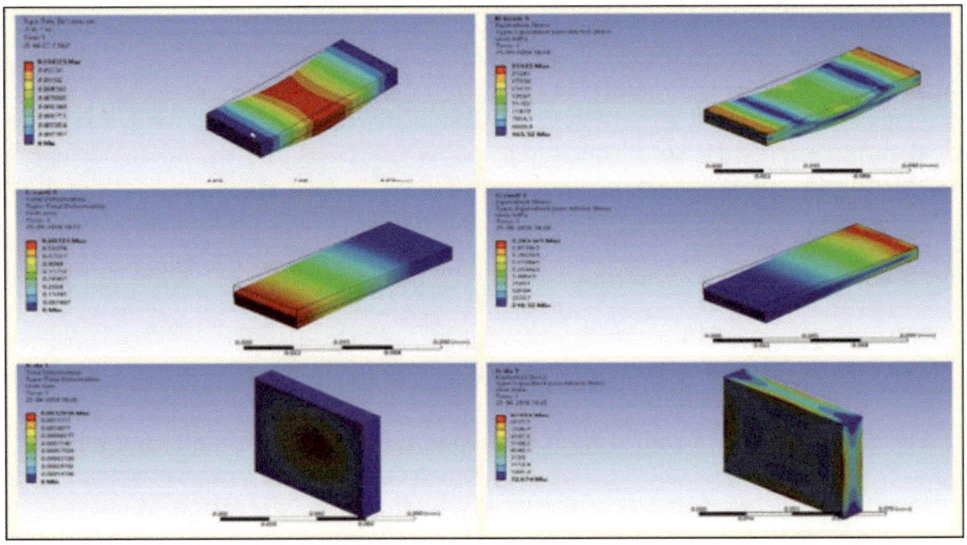

Figure 1.
Total deformation and stress for a beam, cantilever, and a diaphragm.

Device	Total deformation (mm)	Equivalent stress (MPa)
Diaphragm	0.0012956	9244.6
Beam	0.014325	35,125
Cantilever	0.60721	2.2653×10^5

Table 1.
Deformation and stress comparison for mechanical devices.

stress developed on the device. Hence, one of the main criteria is the appropriate structure in order to harvest ambient mechanical vibrations. The mechanical devices that can be used for this purpose are cantilever, beam, and a diaphragm. A diaphragm is a mechanical device which is fixed on four ends, while the cantilever

S. no.	Parameter name	Parameter symbol	Parameter value [units]
1.	Length of substrate layer	L_{sh}	20 [mm]
2.	Width of substrate layer	w_s	8 [mm]
3.	Thickness of substrate layer	t_{sh}	0.04 [mm]
4.	Length of piezoelectric layer	L	20 [mm]
5.	Width of piezoelectric layer	w_p	8 [mm]
6.	Thickness of piezoelectric layer	t_p	0.06 [mm]

Table 2.
Device dimensions.

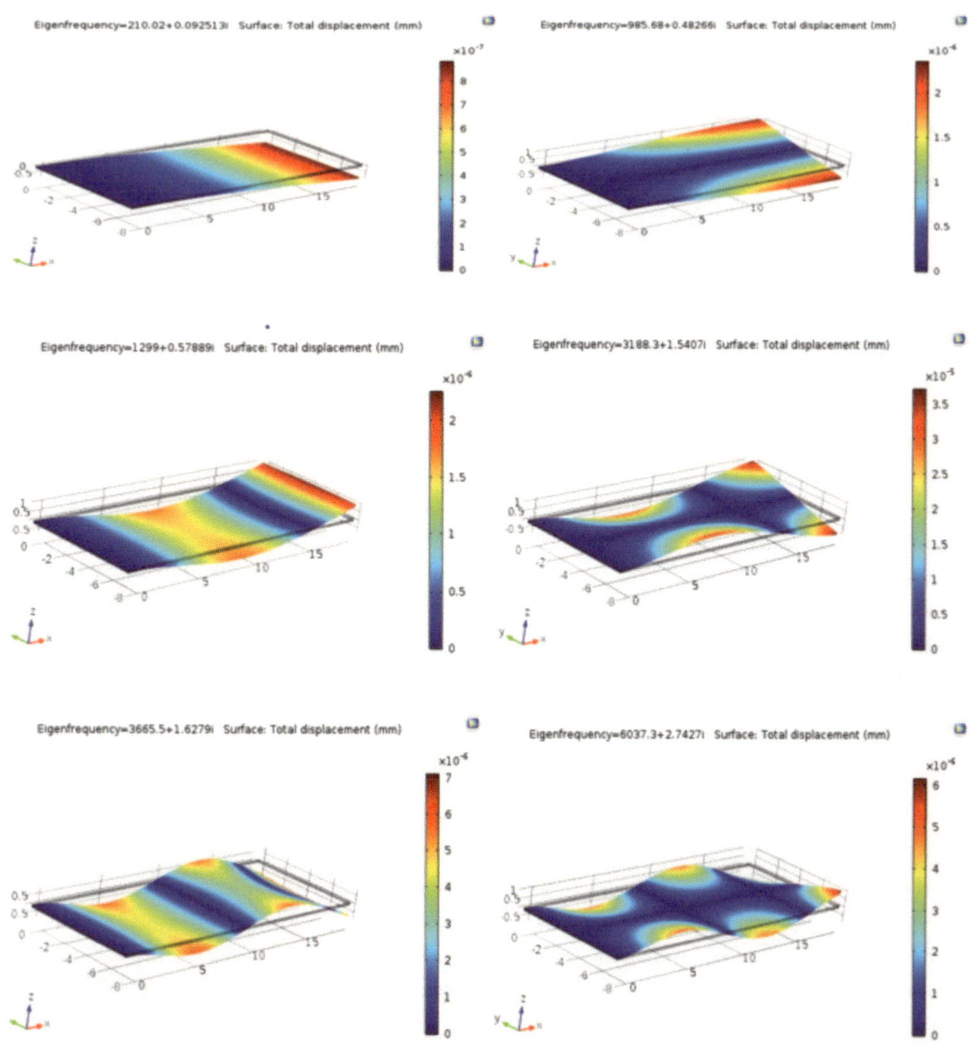

Figure 2.
Six eigenfrequencies of a piezoelectric cantilever bimorph without proof mass.

is a mechanical device which is fixed at one end. In the case of beam, two ends are fixed. The material used in these mechanical devices is silicon. Designs have been simulated in COMSOL Multiphysics.

In order to decide on the appropriate device, all the three devices having the same volume are simulated, and an equal amount of force is applied on each one of them (**Figure 1**). **Table 1** compares the total deformation and equivalent stress values obtained from the above FEM analysis. From **Table 1** it can be inferred that on the application of the same force on cantilever, diaphragm, and beam, a cantilever experiences 42 and 99% more deformation than a beam and a diaphragm, respectively. The stress developed on the cantilever is 25 times more than that on a diaphragm and 7 more than that on a beam. Hence, cantilever structure is mostly used to harvest ambient mechanical vibrations, as for some amount of force; it has the capability to produce a significant stress and displacement.

A piezoelectric cantilever bimorph having piezoelectric layer on top as well as at the bottom is simulated on COMSOL Multiphysics having dimensions as shown in **Table 2**. Eigenfrequencies of the design are plotted in **Figure 2**. Modifications to cantilever by slotting its length have been carried out showing that basic cantilever has the best performance [6, 7]. A broadband piezoelectric energy harvester able to capture ambient vibrations has been designed based on a seesaw cantilever structure [8, 9].

3. Device optimization

To obtain an optimized device, it is essential to select an appropriate substrate and piezoelectric material. Substrate material provides strength and elasticity to the design, while the piezoelectric material performs the conversion of ambient mechanical energy into electrical energy. A substrate material should be able to interact with energy harvesting circuitry as well as provide basic elasticity to the device.

3.1 Selection of substrate material

A cantilever with different substrate material is simulated and its eigenfrequencies are obtained. A comparison of eigenfrequencies for the various designs is shown in **Figure 3**. A cantilever is composed of different substrates having similar dimensions. Aluminum has the lowest first eigenfrequency at around 900 Hz as compared to copper, polyethylene, and structural steel. As the majority of ambient vibrations have lower frequency, hence it is essential for the harvester to have a lower first resonant frequency as then it can match the ambient vibration frequency to produce high power output.

3.2 Selection of piezoelectric material

Piezoelectric materials can be categorized into two types: lead-based and lead-free materials. Lead is a toxic element and has an adverse impact on the environment; hence, its usage is barred in most countries. In this context, scientists are working toward new materials with comparable or better performance than lead-based materials. **Table 3** lists the properties of the piezoelectric materials utilized in this work. Selection of piezoelectric material is equally important for energy harvesting. Piezoelectric material can be lead-based or non-lead-based. Lead-based material has better piezoelectric coefficient than non-lead-based material.

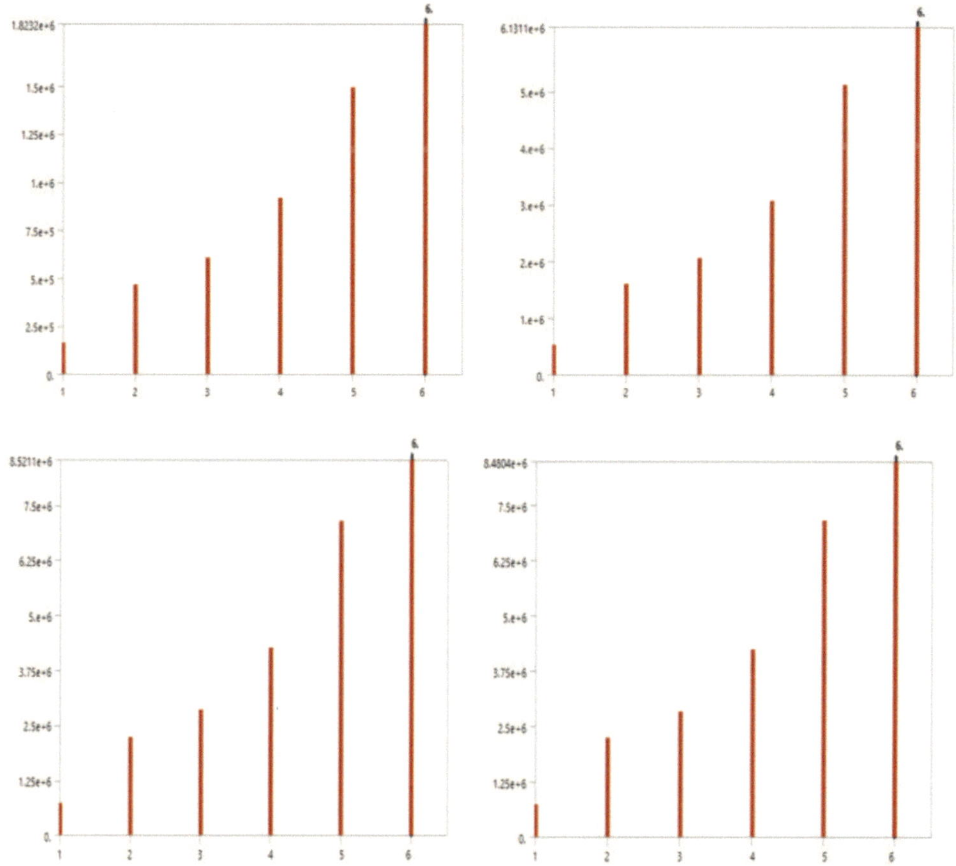

Figure 3.
Six eigenfrequencies of a piezoelectric cantilever bimorph without proof mass using (a) copper, (b) aluminum, (c) polyethylene, and (d) structural steel.

Piezoelectric material	Young's modulus (E) GPa	Density (Kg/m^3)	Relative permittivity	Piezoelectric constants	
				e_{31} (C/m^2)	d_{31} (pC/N)
PZT 5H	127	7500	1433.6	−6.55	−274
PZT 5A	120	7750	826.6	−5.4	−171
PVDF	1.31	1780	7.3	0.0098	−13.6
AlN	149	3300	9	−0.58	−1.72
ZnO	105.3	5680	10.204	−0.56	−5.43

Table 3.
Piezoelectric material properties.

For a dielectric material,

$$D = \varepsilon E \tag{1}$$

For an elastic material,

$$T = sS \tag{2}$$

For a piezoelectric material,

$$D = \varepsilon E + P \quad and \quad T = sS + P \tag{3}$$

The value of P for direct piezoelectric effect is

$$P = d_{31}S \tag{4}$$

And for inverse piezoelectric effect is

$$P = d_{31}E \tag{5}$$

So, constitutive equations for direct and inverse piezoelectric effects are given by

$$D = \varepsilon E + d_{31}S \tag{6}$$

$$T = sS + d_{31}E \tag{7}$$

where D and E are the dielectric displacement and electric field, respectively. T and S are strain and stress on the material. ε and s are electrical permittivity and compliance which is reciprocal to Young's modulus. Piezoelectric cantilever made only with piezoelectric material PZT 5A, PVDF, and zinc oxide is simulated on COMSOL Multiphysics. Results depicted in **Figure 4** show that PVDF has the lowest first eigenfrequency and generates the highest voltage. PVDF has a significantly low Young's modulus, causing it to break easily on application of a very low force. Hence, zinc oxide is selected as piezoelectric material as it is not poisonous and has a significant piezoelectric constant.

Voltage generated in direct piezoelectric effect across a material having t_s thickness is given by

$$\mathrm{V} = \mathrm{t_s E} \tag{8}$$

The charge on a material is given by

$$\mathrm{Q} = \mathrm{AD} \tag{9}$$

Charge is integral of current over a period 't'

$$\mathrm{I} = \mathrm{DAt} \tag{10}$$

Put value of D from (6) to (10):

$$\mathrm{I} = \mathrm{At}(\varepsilon \mathrm{E} + \mathrm{d_{31}S}) \tag{11}$$

Piezoelectric capacitance is given by

$$\mathrm{C} = \frac{\varepsilon \mathrm{A}}{\mathrm{t_s}} \tag{12}$$

Using (8) and (12) in (11) we get:

$$\mathrm{I} = \mathrm{CtV} + \mathrm{Atd_{31}S} \tag{13}$$

For open circuit condition I = 0. Hence, (13) solves to

$$\mathrm{V} = -\frac{Ad_{31}S}{\mathrm{C}} \tag{14}$$

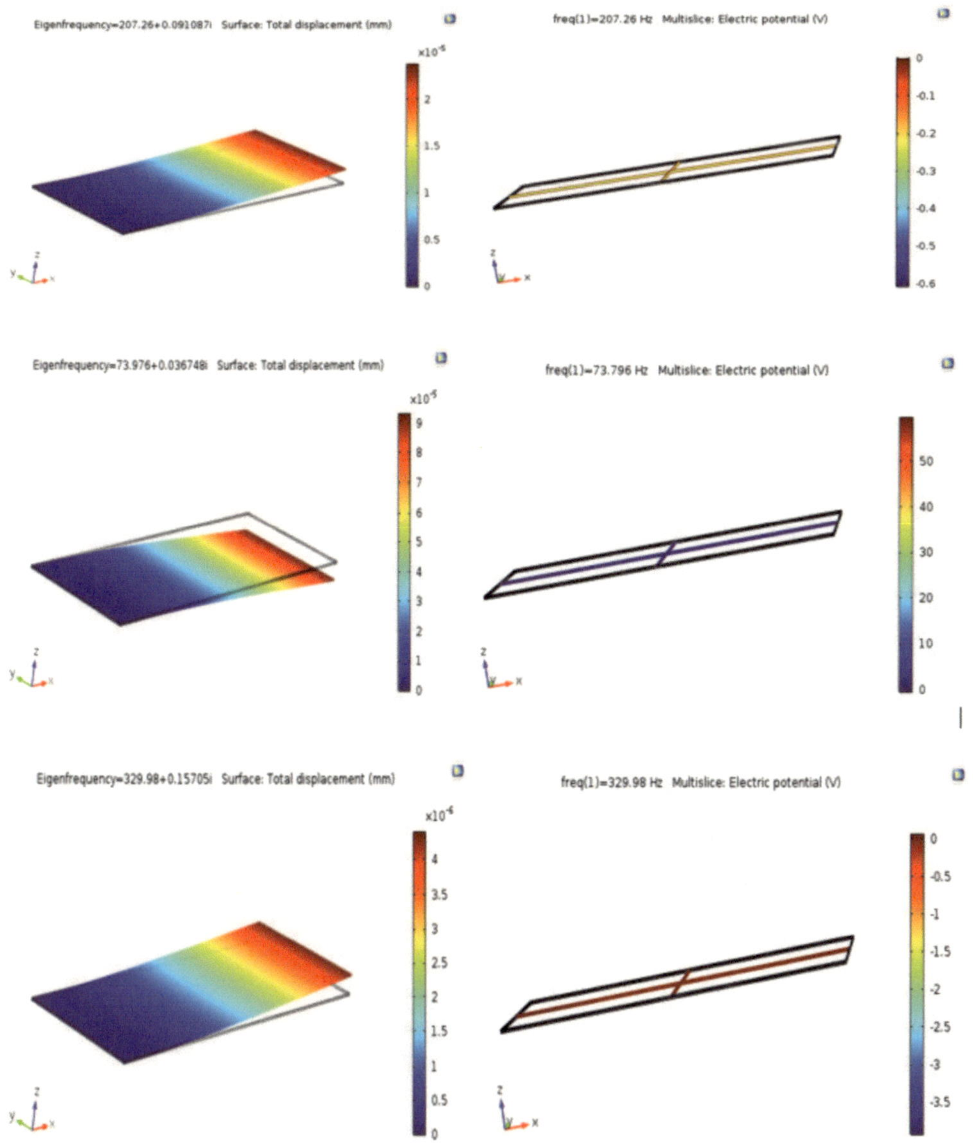

Figure 4.
Eigenfrequencies of a piezoelectric cantilever bimorph without proof mass using (a) PZT 5A, (b) PVDF, and (c) zinc oxide.

Putting the value of C back from (12), we get

$$V = -\frac{d_{31} S\, t_p}{\varepsilon} \tag{15}$$

Hence, on application of S stress on a piezoelectric material of thickness t_p having piezoelectric constant d_{31} and permittivity ε, the voltage generated is given by Eq. (15).

3.3 Effect of proof mass

Proof mass is added to a piezoelectric cantilever to reduce its eigenfrequency. The dimensions of the proof mass added to the cantilever are given in **Table 4**.

S. no.	Parameter name	Parameter symbol	Parameter value [units]
1.	Length of proof mass	l_m	10 [mm]
2.	Width of proof mass	w_m	8 [mm]
3.	Height of proof mass	t_m	5 [mm]

Table 4.
Dimensions of proof mass.

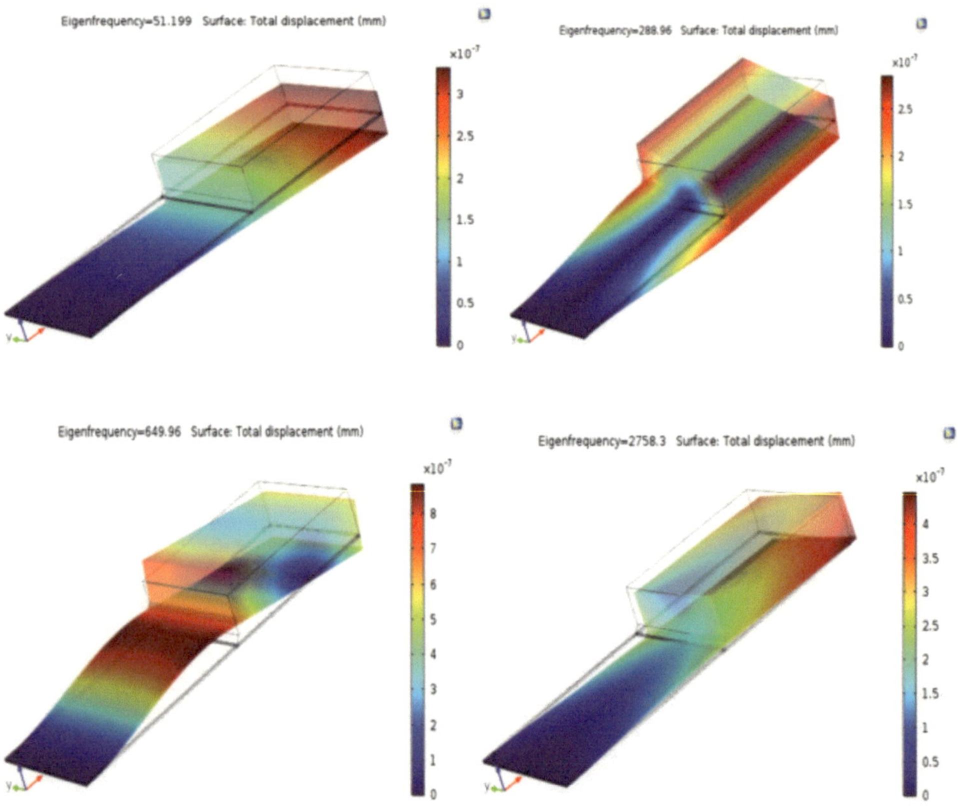

Figure 5.
First four eigenfrequencies of a piezoelectric cantilever bimorph with proof mass.

When the proof mass is added, the first eigenfrequency reduces to 51.3 Hz as shown in **Figure 5**. The cantilever becomes bulkier. The material of proof mass is the same as that of the substrate. Proof mass also increases the overall mass of the device and hence, the displacement of the device from the neutral position.

4. Results and discussion

4.1 Frequency response

The frequency response of the piezoelectric cantilever obtained is plotted by varying the input vibration frequency (**Figures 6–8**). The frequency response depicts a peak when the frequency of vibration matches the first eigenfrequency.

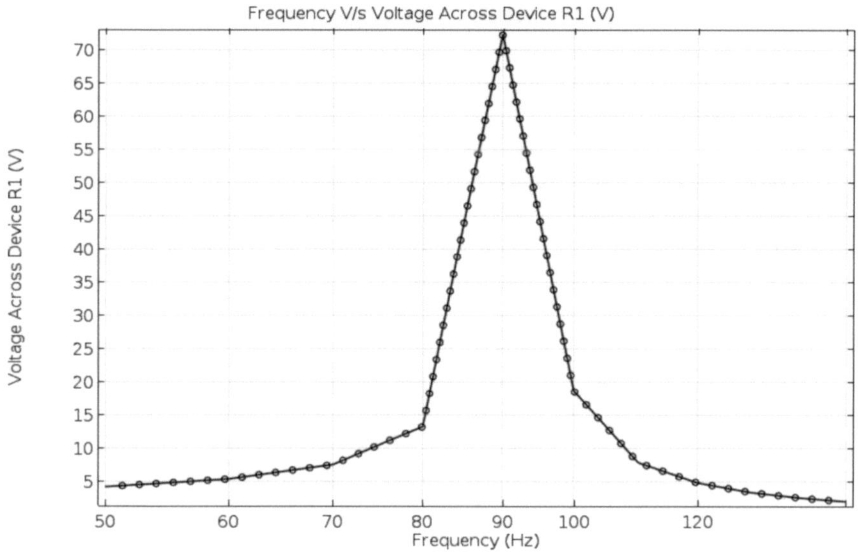

Figure 6.
Voltage generated on a frequency response of a piezoelectric cantilever bimorph.

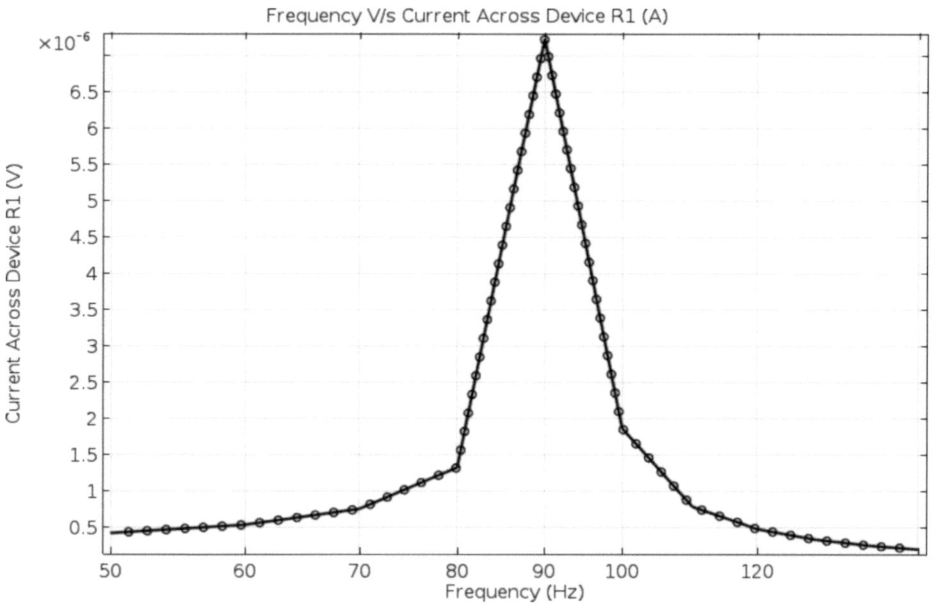

Figure 7.
Current generated on a frequency response of a piezoelectric cantilever bimorph.

The cantilever has two piezoelectric layer of zinc oxide and a substrate layer of aluminum. The cantilever energy harvester can generate a power of 0.47 mW across a 4 Mohm resistor. The current value is around 6.8 μA. Generated power can be increased by integrating the harvester with energy harvesting circuitry [10, 11].

4.2 Electromechanical analysis

A significant part of the piezoelectric energy harvesting process is mechanical to electrical conversion. High power is generated on high amount of stress on the

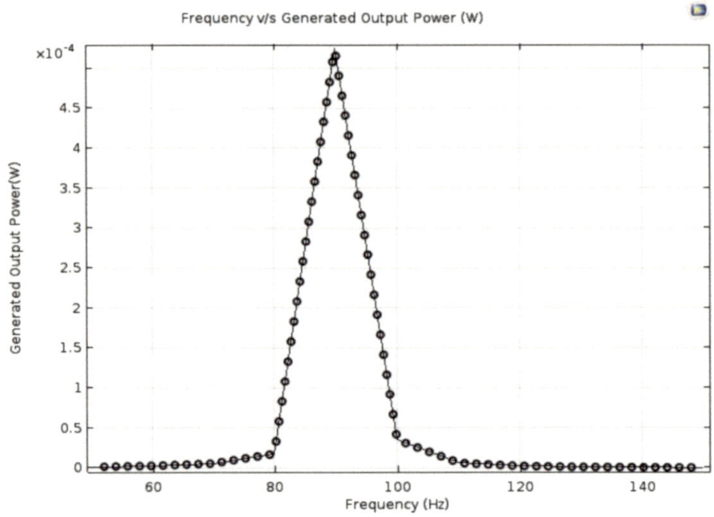

Figure 8.
Power generated on a frequency response of a piezoelectric cantilever bimorph.

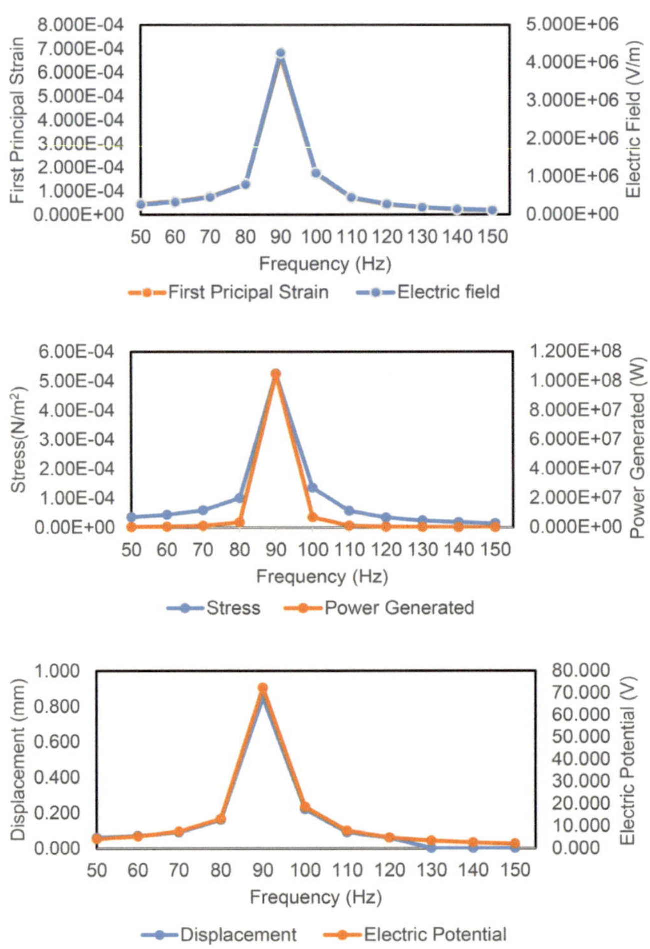

Figure 9.
Electromechanical conversion on a frequency response of a piezoelectric cantilever bimorph energy harvester for strain to electric field, stress to power generated, and displacement to electric field.

piezoelectric layer. Electromechanical coupling displays the conversion of stress on the piezo layer and equal conversion of electrical power by those layers. The more the stress, the larger is the displacement of the device which increases the quantity of accumulated charges and the higher is the potential developed on the layer. **Figure 9** shows electromechanical conversion from first principal strain to electric field, displacement to electric potential, and stress to generated output power.

A peak strain of 6.712×10^{-4} resulted in a peak electric field of 4.26×10^{6} V/m with both occurring near the first resonant frequency. A peak stress of 1.05×10^{8} Pa resulted in power generation of around 0.5 mW with both occurring near the first eigenfrequency. A peak displacement of 0.85 mm resulted in peak electric potential of 72.394 V, with both occurring near the first eigenfrequency.

5. Conclusions

A piezoelectric cantilever bimorph is simulated in this chapter. Optimizing the designing from the selection of substrate and piezoelectric material is described. A cantilever with aluminum substrate and zinc oxide is designed with a proof mass. Analytical model of generated voltage by the piezoelectric cantilever energy harvester is also described. Addition of proof mass reduces the resonant frequency of the device. Power frequency response functions and electromechanical analysis of the resultant structure are able to generate 0.47 mW power having 6.8 μA current at 1 g input acceleration.

Author details

Prateek Asthana* and Gargi Khanna
National Institute of Technology, Hamirpur, Himachal Pradesh, India

*Address all correspondence to: prateekasthana1989@gmail.com

References

[1] Marzencki M, Ammar Y, Basrour S. Integrated power harvesting system including a MEMS generator and a power management circuit. Sensors and Actuators A: Physical. 2008;**145**:363-370

[2] Anton SR, Sodano HA. A review of power harvesting using piezoelectric materials (2003–2006). Smart Materials and Structures. 2007;**16**(3):R1

[3] Sodano HA, Inman DJ, Park G. Comparison of piezoelectric energy harvesting devices for recharging batteries. Journal of Intelligent Material Systems and Structures. 2005;**16**(10): 799-807

[4] Minami Y, Nakamachi E. Development of enhanced piezoelectric energy harvester induced by human motion. Conference Proceedings: Annual International Conference of the IEEE Engineering in Medicine and Biology Society. 2012;**2012**:1627-1630

[5] Hwang G-T et al. Self-powered cardiac pacemaker enabled by flexible single crystalline PMN-PT piezoelectric energy harvester. Advanced Materials. 2014;**26**(28):4880-4887

[6] Asthana P, Khanna G. Design and analysis of slotted cantilever structure for piezoelectric energy harvesting. In: 2017 4th International Conference on Signal Processing, Computing and Control (ISPCC); 2017. pp. 387-390

[7] Asthana P, Dwivedi A, Khanna G. Equivalent Circuit Modelling for Unimorph and Bimorph Piezoelectric Energy Harvester. In: International Conference on Advanced Informatics for Computing Research. Singapore: Springer; 14 Jul 2018. pp. 39-49

[8] Asthana P, Khanna G. Finite-element modeling of piezoelectric energy harvesters using lead-based and lead-free materials for voltage generation. Journal of Asian Ceramic Societies. 2018;**6**(4):394-400

[9] Asthana P, Khanna G. A broadband piezoelectric energy harvester for IoT based applications. Microelectronics Journal. 2019;**93**:104635

[10] Asthana P, Dwivedi A, Khanna G. Finite Element Modeling of a Wideband Piezoelectric Energy Harvester for Ambient Vibration Extraction. In: Advances in VLSI, Communication, and Signal Processing. Singapore: Springer; 2020. pp. 549-556

[11] Khanna G, Asthana P. A comparative study of circuit for piezo-electric energy harvesting. In: 2016 3rd International Conference on Computing for Sustainable Global Development (INDIACom); 2016. pp. 1689-1694

Chapter 4

Ferroelectric Polymer PVDF-Based Nanogenerator

Jeongjae Ryu, Seongmun Eom, Panpan Li, Chi Hao Liow and Seungbum Hong

Abstract

This chapter deals with the development of ferroelectric polymer polyvinylidene fluoride (PVDF)-based nanogenerators. Due to its inherent flexibility, PVDF has been studied for application in nanogenerators. We first introduce PVDF and its copolymers, and briefly discuss their properties. Then, we discuss fabrication methods, including solution casting, spin coating, template-assisted method, electrospinning, thermal drawing, and dip coating. Using these methods, a wide variety of ferroelectric polymer structures can be fabricated. In addition to the performance enhancements provided by fabrication methods, the performance of PVDF-based nanogenerators has been improved by incorporating fillers that can alter the factors affecting the performance. Next, we review energy sources that can be exploited by PVDF-based nanogenerators to harvest electricity. The abundant energy sources in the environment include sound, wind flow, and thermal fluctuation. Finally, we discuss implantable PVDF-based nanogenerators. Another advantage of PVDF is its biocompatibility, which enables implantable nanogenerators. We believe that this chapter can also be helpful to researchers who study sensors and actuators as well as nanogenerators.

Keywords: PVDF, nanogenerator, biocompatibility, flexibility, ferroelectric material

1. Introduction

All electronic devices need electrical energy to operate. Although fossil fuels have been the primary sources of that electrical energy to date, alternatives are emerging. This is particularly important given the ongoing proliferation of portable devices. For example, as "big data" has become increasingly important for monitoring structures, healthcare services, smart cites, and so on, there has been an explosion of sensors to collect that information [1–3]. These sensors can even be located in the human body, beneath the human skin, and inside personal wearable devices [4–7]. In such cases, batteries cannot be easily charged or repeatedly replaced. To meet the energy requirements of remote or inaccessible application like these, one of the most promising alternatives to conventional batteries is a nanogenerator, which can convert mechanical energy or thermal fluctuations in the ambient environment into electricity.

Polyvinylidene fluoride (PVDF) and its copolymers, ferroelectric polymers, are ideal candidates for use in nanogenerators. Their unique properties include their high flexibility, lightness, chemical stability, and relatively simple manufacturing

 IntechOpen

Materials	Microstructures	Structures	Open-circuit voltage	Short-circuit current	Power	References
P(VDF-TrFE)	—	Flat film	7 V	58 nA	—	[12]
P(VDF-TrFE)	—	Curved film	120 V	700 μA	3.9 mW·cm^{-2}	[13]
PVDF	—	Fabric	14 V @ 0.1 MPa	29.8 μA @ 0.1 MPa	5.1 μW·m^{-2}	[14]
PVDF/Ba($Ti_{0.9}Zr_{0.1}$)O_3	Nanocubes	Flat film	11.99 V @ 11 N	1.36 μA @ 11 N		[15]
PVDF/SnO_2	Nanosheets	Flat film	42 V	6.25 μA·cm^{-2}	4900 W·m^{-3}	[16]
PVDF/ZnO	Nanoparticles	Flat film	24.5 V @ 28 N	1.7 μA @ 28 N		[17]
PVDF/ZnO	Nanowires	Flat film	6.9 V	0.96 μA	6.624 μW	[18]
PVDF/AlO-rGO	Nanoparticles	Flat film	36 V @ 31.19 kPa	0.8 μA @ 31.19 kPa	27.97 μW	[19]
PVDF/$BaTiO_3$	Nanoparticles	Flat film	10 $V_{peak\text{-}peak}$ @ 2 N	2.5 $\mu A_{peak\text{-}peak}$ @ 2 N	5.8 μW	[20]
PVDF/$BaTiO_3$	Nanowires	Flat film	14 V	4 μA	1.5 μW	[21]
PVDF/NKNS-LT-BZ	Nanoparticles	Flat film	18 V @ 50 N	2.6 μA @ 50 N	—	[22]
PVDF/NiO@SiO_2	Nanoparticles	Flat film	53 V @ 0.3 MPa	0.3 μA·cm^{-2} @ 0.3 MPa	685 W·m^{-3}	[23]
PVDF	—	Electrospun membrane	48 V @ 8.3 kPa	6 μA @ 8.3 kPa	51 μW	[24]
PVDF/ZnO	Nanorods	Electrospun membrane	85 V	2.2 μA	—	[25]

Table 1.
Comparison of the PVDF-based nanogenerator performance. The experimental conditions such as applied force, operation mode, area of nanogenerator, and load resistance were different in the literature.

process [8–11]. The limitation of PVDF-based nanogenerators has been their low power generation capacity. Hence, many research groups have studied to enhance the performance of nanogenerators, by utilizing various fabrication methods, and structures, by incorporating fillers (**Table 1**).

In this chapter, we discuss different fabrication techniques and developments of PVDF-based nanogenerators in detail. This book chapter is organized as follows. In Section 2, we introduce PVDF and its copolymers with their properties. In Section 3, we focus on fabrication methods used to prepare PVDF-based nanogenerators. In Section 4, we briefly cover conventional PVDF film-based nanogenerators. In Section 5, we review composite-based nanogenerators and how certain factors affect their performance of the nanogenerators. In Section 6, we introduce the energy sources that can be harvested by the nanogenerators. In Section 7, we review the biocompatibility of PVDF and related works.

2. PVDF and its copolymers

Semicrystalline PVDF has at least four crystalline modifications: α, β, γ, and δ [26]. Generally, the α-phase is the most stable crystal phase of PVDF in ambient conditions. The conformation of α-PVDF, displayed in **Figure 1a**, is trans-gauche-trans-gauche (TGTG'). The α-phase is nonpolar because of its centrosymmetric symmetry and can be easily obtained from melt crystallization at atmospheric pressure. On the other hand, the β-phase is classified as a ferroelectric and exhibits the largest remnant polarization of ~ 13 $\mu C/cm^2$ among the phases. Ferroelectricity in the β-PVDF is directly correlated to its macroscopic dipole moment. In the β-phase,

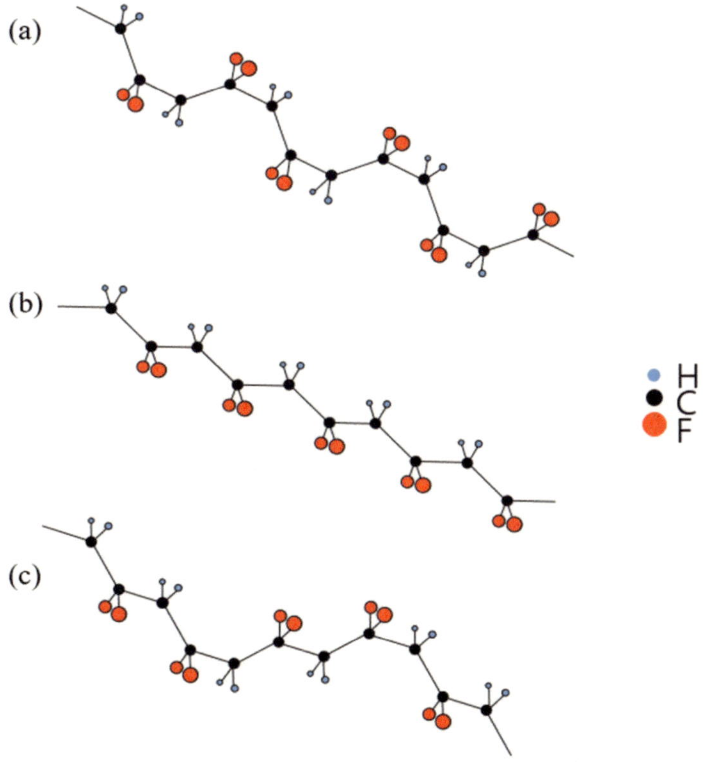

Figure 1.
Configuration in the unit cell of (a) the α-phase, (b) the β phase, and (c) the γ-phase.

the conformation is all trans (TTTT) as shown in **Figure 1b**. The transition from the PVDF α-phase to the β-phase can be induced by stretching the polymer. The γ-phase is also polar and its conformation is trans-trans-trans-gauche-trans-trans-trans-gauche (T_3GT_3G) as shown in **Figure 1c**. The γ-phase is also ferroelectric. However, the γ-phase is less frequently observed because it requires extreme temperature control and high pressures to develop. The δ-phase has the same configuration as the α-phase. The difference is that the δ-phase is a ferroelectric. Crystallization of the δ-phase can be achieved by electroforming from the bulk α-PVDF in a high electric field of about 170 MV/m. Recently, M. Li *et al.* demonstrated that a δ-PVDF thin film can be made at elevated temperatures by applying a short pulse of about 250 MV/m [27]. They confirmed the existence of the δ-phase via combined GI-XRD and FTIR measurements on pristine and electroformed capacitors.

Due to its electroactive properties, high β-phase content is desirable for applications. This can be achieved physically or chemically. The most common chemical derivatives of PVDF available are polyvinylidene fluoride-trifluoroethylene (P(VDF-TrFE)) and polyvinylidene fluoride-hexafluoropropene (P(VDF-HFP)). Despite its high cost, P(VDF-TrFE) is often preferred over PVDF. The main advantage of P(VDF-TrFE) is that the β-phase can easily develop without mechanical stretching or incorporating fillers. The mechanism behind the development of the β-phase is based on the introduction of additional fluorine atoms within a certain amount, which allows steric hindrance to occur. P(VDF-HFP) has received a lot of attention because of its extremely high electrostrictive response [28]. As with PVDF, the β-phase in P(VDF-HFP) can be obtained by mechanical stretching [29]. Another way to develop the β-phase in P(VDF-HFP) is casting from solution in dimethylformamide (DMF) [30].

Since the discovery of piezoelectricity in PVDF [31], the mechanism behind the piezoelectric response has been a subject of debate. Recently, L. Katsouras *et al.* identified the piezoelectric effect in semicrystalline PVDF and its copolymers via *in-situ* dynamic X-ray diffraction, from measurements with P(VDF-TrFE) capacitors [32]. The piezoelectric effect is attributed to the change in lattice constant produced by the electromechanical coupling between the intermixed crystalline and amorphous regions as well as the electrostrictive response of the crystalline part. Both the electrostrictive Q_{33} and the additional, $d_{coupling}$, are negative; hence, the piezoelectric coefficient, d_{33}, of P(VDF-TrFE) is negative. Understanding the fundamental basis of piezoelectricity in ferroelectric polymers is essential to further technological innovations.

In order to identify the phases of PVDF, analysis techniques based on electromagnetic radiations such as X-ray and infrared have been widely used. The crystalline structure in PVDF and its copolymers can be confirmed via X-ray diffraction (XRD). In general, the nonpolar α-phase and the polar β- and γ-phases of PVDF appear in XRD patterns (**Figure 2a**) [33]. There are peaks at 18.4, 19.9, and 26.6° corresponding to (020), (110), and (021) reflections of the monoclinic α-phase, respectively. The peak at 20.6° is associated with the crystalline (200) and (110) of the orthorhombic β-phase. For the γ-phase, the dominant peaks appear at 18.5 and 20.2° corresponding to (020) and (110), respectively. In the case of P(VDF-TrFE), the peak corresponding to the (200) and (110) planes of the β-phase crystalline phase is represented at 19.7°. The position and width of peaks can change depending on experimental conditions and the ratio between VDF and TrFE. Therefore, the diffraction patterns can be differently observed in the literature.

The ferroelectric phase formation of PVDF and its copolymers have been also confirmed using Fourier transform infrared spectroscopy (FTIR). However, there exists a conflict on spectrum peaks corresponding the phases. Recently, X. Cai *et al.* documented reports on the vibrational bands of PVDF materials via FTIR to shed

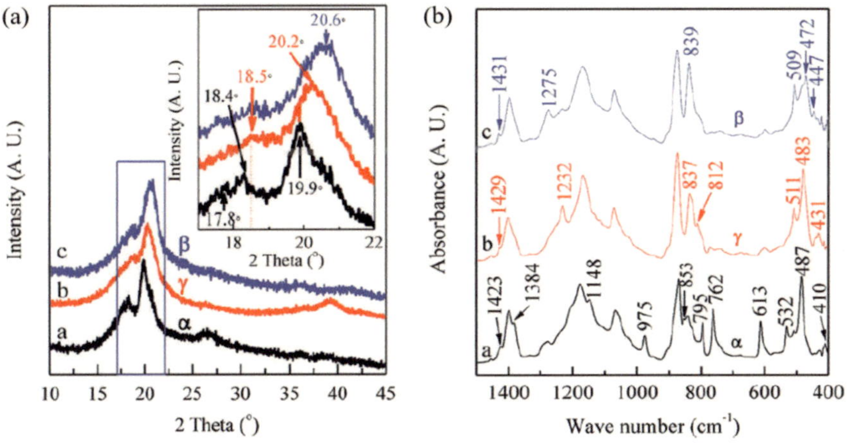

Figure 2.
XRD patterns (A) and FTIR spectra (B) of PVDF membranes [33].

light on this issue [33]. Furthermore, they analyzed the FTIR results and suggested the relative fraction of the β- and γ-phase calculations in terms of crystalline components as shown in **Figure 2b**. Since the absorption band at 840 cm^{-1} corresponds to the β, γ, or both phases, the fraction of the β- and γ-phases can be calculated as follows:

$$F_{EA} = \frac{I_{EA}}{\left(\frac{K_{840}}{K_{763}}\right) I_{763} + I_{EA}} \times 100\% \quad (1)$$

where I_{EA} and I_{763} are the absorbance at 840 and 763 cm^{-1}, respectively, and K_{840} and K_{763} are the absorption coefficients at the respective wave numbers, whose values are 7.7×10^4 and 6.1×10^4 $cm^2\ mol^{-1}$, respectively. If PVDF materials contain both phases, another method is needed to quantify individual β- and γ-phases. The quantification of β- and γ-phases can be demonstrated as follows:

$$F(\beta) = F_{EA} \times \left(\frac{\Delta H_{\beta'}}{\Delta H_{\beta'} + \Delta H_{\gamma'}}\right) \times 100\% \quad (2)$$

$$F(\gamma) = F_{EA} \times \left(\frac{\Delta H_{\gamma'}}{\Delta H_{\beta'} + \Delta H_{\gamma'}}\right) \times 100\% \quad (3)$$

where $\Delta H_{\beta'}$ and $\Delta H_{\gamma'}$ are the height differences (absorbance differences) between the peak around at 1275 cm^{-1} and the nearest valley around at 1260 cm^{-1}, and the peak around 1234 cm^{-1} and the nearest around at 1225 cm^{-1}, respectively.

3. Fabrication techniques

PVDF and its copolymer-based nanogenerators consist of a layer of PVDF or its copolymers sandwiched between two electrodes, like a capacitor. As previously noted, PVDF and its copolymer-based nanogenerators have advantages over piezoelectric ceramics-based nanogenerators, because of their flexibility and high piezoelectric voltage coefficient (g_{33}). One important consideration is the choice of a suitable technique for the fabrication of ferroelectric polymer-based nanogenerators. A great variety of methods to make ferroelectric polymer-based nanogenerators has been developed for a few decades: solution casting, spin coating, template-assisted method, electrospinning, thermal drawing, and dip coating, as

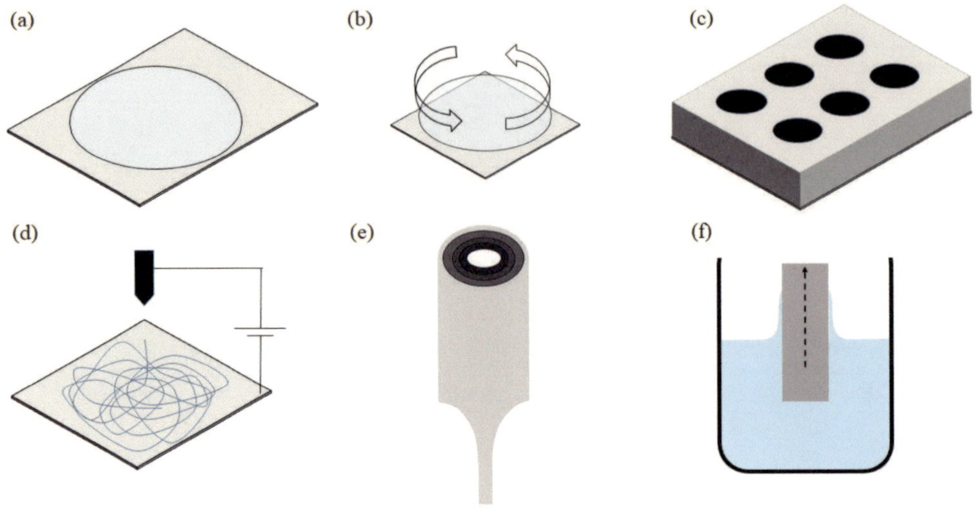

Figure 3.
Schematic drawing of fabrication methods of PVDF-based nanogenerator. (a) Solution casting, (b) spin coating, (c) template-assisted method, (d) electrospinning, (e) thermal drawing, (f) dip coating.

described in **Figure 3** [9, 12, 34–38]. The particular fabrication technique can affect the crystallinity and phase of ferroelectric polymers, influencing their electroactive and mechanical properties. This section covers conventional and recently developed fabrication techniques involving ferroelectric polymers in greater detail.

3.1 Solution casting

Usually, PVDF and its copolymer-based nanogenerators are prepared by solution casting because it is a simple and low-cost process and allows large-scale production. The casting method is also suitable for preparing composites of fillers and a polymer matrix because it is relatively insensitive to the viscosity and density of solution. As a result, solution casting has been used in most research on composite-based nanogenerators [16, 18, 22, 23, 39]. The thickness of the film can be controlled by the type of organic solvents employed and the concentration and amount of the ferroelectric polymer solution. Since evaporation speed can differ depending on the organic solvents used, the morphology of the ferroelectric polymer films can also be affected [40].

3.2 Spin coating

Spin coating is a universally used technique to fabricate uniform thin polymer films using the centrifugal forces induced by a spinning substrate. The thickness of the films can be controlled by a few parameters, such as solution concentration, viscosity, spin speed, and spin time. Generally, a low concentration polymer solution (<10%) is used for uniformity. In addition, the thickness of spin-coated ferroelectric polymer films can be controlled by the number of successive spin coatings.

Spin coating has been successfully used to prepare nanogenerators based on thin P(VDF-TrFE) films. For example, Z. Pi *et al.* reported a flexible nanogenerator based on P(VDF-TrFE) thin film produced by spin coating. A 2% concentration of P(VDF-TrFE) copolymer solution was repeatedly spin coated on the substrate to reach the required thickness [12]. In another study, Lee *et al.* successfully developed a nanogenerator based on micropatterned P(VDF-TrFE) polymer (thickness ≈7 μm) [41].

They evaluated the improved stretchability, mechanical durability, and robustness of the line-based micropatterned nanogenerator, compared to a flat (nonpatterned) nanogenerator.

3.3 Template-assisted method

Nanoconfinement involves confining dimensional geometries in a nanosized region. Typically, the template-assisted method from a melt or solution has been employed to fabricate one-dimensional nanostructures [8, 9]. In the template-assisted method, a polymer melt or solution is used to wet the surface of a porous template. It permeates into the pore of the template due to the high surface energy of the template walls. Finally, polymer nanotubes or nanowires form with a uniform size distribution. This technique has been reported to improve the performance of polymer-based nanogenerators. For example, V. Bhavanasi *et al.* prepared P(VDF-TrFE) nanotubes using a template-assisted method, with a P(VDF-TrFE) melt and anodic aluminum oxide (AAO) templates with a pore diameter of 200 nm [42]. The resulting P(VDF-TrFE) nanotubes demonstrated high power generation under dynamic compression pressure compared to that of P(VDF-TrFE) films. This improved performance was attributed to the enhanced piezoelectric coefficients, strain confinement in the 1D geometry, and improved surface area.

The nanoconfinement via templates induces ferroelectric polymer nanowires to retain preferential orientation with the b-axis parallel to the long axis of the template. This orientation leads to a vertical direction of the polarization parallel to the channel axis. Y. Calahorra *et al.* investigated the electromechanical interactions in individual P(VDF-TrFE) nanowires via piezoresponse force microscopy (PFM) [43]. Their results indicate that P(VDF-TrFE) nanowires grown via template-assisted methods possess piezoelectric properties without electrical poling process, which is called as "self-poled." However, the d_{33} coefficient of nanowires without poling was lower than that of nanowires with poling. After tip-induced electrical poling, the d_{33} of the nanowires increased from 7.7 pm/V to 21.6 pm/V. This phenomenon can be attributed to the direction of polarization parallel to the substrate before electrical poling.

When a template-assisted method is used, in-situ poling can be applied at the same time. This is known as the template-assisted electricity-grown method. X. Chen *et al.* fabricated vertically well-aligned P(VDF-TrFE) nanowire arrays with a preferential polarization orientation [44]. The contents of the β-phase in P(VDF-TrFE) nanowires and bulk films were analyzed via Fourier transform infrared (FTIR) measurement. The percent content of β-phase in the P(VDF-TrFE) nanowires increased to 83.5%. This value was higher than the bulk films (75.2%). Without further poling, the P(VDF-TrFE) nanowire array delivered an output voltage about 9 times higher than conventional spin-coated bulk films.

3.4 Electrospinning

Electrospinning is another common method used to fabricate β-phase ferroelectric fibers without further mechanical stretching or electrical poling processes [45]. The electrospinning technique utilizes an electrical force to obtain fibers. The basic principle of conventional electrospinning is as follows: When a DC electric field is applied between the spinneret (metallic needle) and the collector (grounded conductor), a conical object called a Taylor cone forms at the tip of the needle. When the electric field rises above a threshold value, the resulting electrostatic force can overcome the surface tension and viscous force of the polymer solution. An electrified liquid jet ejected from the nozzle undergoes a stretching and whipping process and then splits into threads with diameters ranging from hundreds of micrometers to tens of nanometers.

In 2011, D. Mandal *et al*. prepared P(VDF-TrFE) nanofiber webs via electrospinning [37]. They found evidence of the preferential orientation of CF_2 dipoles in the P(VDF-TrFE) nanofiber web via analyses using polarized FTIR and piezoelectric signal. They also demonstrated that output current signal could be enhanced by stacking several tens of layers of the electrospun web with the same polarization.

Unlike conventional electrospinning, near-field electrospinning using a short needle-to-collector distance enables superior location control, to produce orderly nanofiber patterns over large areas [46]. Notably, C. Chang *et al*. utilized a direct-write technique by means of near-field electrospinning (NFES) to produce and place piezoelectric PVDF nanofibers on working substrates [47]. They measured the piezoelectric response of single PVDF nanofibers to calculate their energy conversion efficiency (ECE). The ECE was defined as

$$ECE = \frac{\textit{Electric energy generated}}{\textit{Mechancial energy applied}} = \kappa^2 \tag{4}$$

where κ is the electromechanical coupling coefficient of the piezoelectric material. The electric energy W_e was estimated by integrating the product of the output voltage and current of the PVDF nanogenerator when stretched. The elastic strain energy was estimated using

$$W_s = \tfrac{1}{2} E A \varepsilon^2 L_0 \tag{5}$$

where E is the Young's modulus of the material, A is the cross-sectional area, ε is the strain applied to the material, and L_0 is the length of the material. It was found that the energy conversion of the nanofiber-based nanogenerator (average 12.5%) was higher than experimental PVDF films (0.5–4%) and conventional PVDF thin films (0.5–2.6%).

3.5 Thermal drawing

Another interesting method to make ferroelectric polymer fibers is thermal drawing, as reported by S. Egusa *et al* [38]. The advantage of thermal drawing is that all components (piezoelectric layer, electrode layers, and wrapping layer) are extracted a preform made of poly(vinylidene fluoride) with metal electrodes and an insulating polymer, making the process simpler compared to other methods. However, they encountered some challenges. During thermal drawing, there is a susceptibility to break up and mixing due to flow instabilities and variation in layer thickness. To address such challenges, they used carbon-loaded poly(carbonate) (CPC) and P(VDF-TrFE). CPC has high viscosity (10^5–10^6 Pa s) at the draw temperature and thus can be used to confine the low-viscosity crystalline piezoelectric layer and ensure adequate conductivity (1–10^4 Ω m). In the case of P(VDF-TrFE), the β-phase can form without any mechanical stress compared to PVDF. Therefore, they were able to make multimaterial piezoelectric fibers with a uniform diameter using these two materials. They also demonstrated the benefits of drawing ferroelectric polymer fibers for application in acoustic transducers and capacitors [38, 48]. However, they did not mention the scavenging ability of P(VDF-TrFE) fiber prepared by thermal drawing.

Due to flexibility of the thermal drawing, it has been successfully demonstrated for fabricating composite fibers such as $BaTiO_3$-PVDF, $Pb(Zr,Ti)O_3$ (PZT)-PVDF, and carbon nanotube (CNT)-PVDF [49]. In that study, the piezoelectric performance of the fabricated fibers was systematically compared while bending and releasing. For a $BaTiO_3$-PVDF fiber, the generated open-circuit voltage and short-circuit current were 1.4 V and 0.8 nA, respectively. For a PZT-PVDF fiber, the corresponding values were 6 V and 4 nA. A CNT-PVDF fiber generated 3 V and 1.2 nA. These results were attributed to the high piezoelectric coefficient ($BaTiO_3$ and PZT) or the induced β-crystallization (CNT).

3.6 Dip coating

Dip coating is one of the most effective methods of depositing ferroelectric polymers on a 3D substrate. For example, D. Kim *et al.* developed a three-dimensional spring-type nanogenerator using a dip-coating and multidirectional electrode deposition [35]. The nanogenerator had a bi-layered structure with a top electrode and a P(VDF-TrFE) layer on a conventional spring that had two roles. It served as both the core electrode and the mechanical substrate for the ferroelectric polymer. In another study, J. Ryu *et al.* fabricated a P(VDF-TrFE)/polydimethylsiloxane (PDMS) stretchable hollow fiber using dip coating [36]. To fabricate the fiber, a cylindrical carbon rod was used as a template, onto which each of the layers was sequentially coated. The diameters of the P(VDF-TrFE)/PDMS layer and PDMS layer were controlled by varying the withdrawal speed during dip coating.

4. Nanogenerators based on pure PVDF films

The typical type of nanogenerator is based on ferroelectric polymer films without any further processing or incorporation of fillers. Conventional PVDF film-based nanogenerators can be simply fabricated and easily stacked to improve performance. In addition, the flexible nanogenerators can be designed using various structures. For example, J. Zhao and Z. You fabricated a nanogenerator based on specially designed sandwich structure that was compatible with a shoe [50]. The structure was composed of multilayered PVDF films and two wavy surfaces of a movable upper plate and a lower plate. The PVDF film layers were connected in parallel for high output current. Due to the structural design and stacked PVDF films, the nanogenerator provided an average output power of 1 mW while walking at a frequency of ~1 Hz. In another study, W.-S. Jung *et al.* prepared a curved piezoelectric nanogenerator consisting of multilayered PVDF films and a polyimide substrate [13]. They also connected all PVDF films in parallel to increase the output current. The curved structure enhanced the output power by distributing effectively applied stress. The authors integrated the nanogenerator into a shoe-insole and to a watch strap and attached them to long-sleeved shirts, and then evaluated the PVDF-based nanogenerators as a power source for wearable sensors and biomedical devices.

5. Nanogenerators based on composites

In order to enhance the electrical and mechanical properties of ferroelectric materials, many research groups have employed the concept of composites, typically made of a polymer and a ceramic. Composite is a common approach in piezoelectric-based applications. Ultrasonic transducer is a typical example. The composites are used for impedance matching. Since piezoelectric ceramics like PZT, $BaTiO_3$, and $Pb(Mg_{1/3}Nb_{2/3})O_3$-$PbTiO_3$ (PMN-PT) have a relatively high acoustic impedance compared to water or human tissue, impedance matching is required. Such matching results in broad bandwidth and increased sensitivity.

Composites can also be utilized for nanogenerators based on PVDF and its copolymers to boost their performance. The outputs of nanogenerators depend on a variety of factors, which include the piezoelectric coefficient, elastic modulus, and dielectric constant. The output voltage generated by nanogenerators can be expressed as

$$V_{out} = \int g_{33}\varepsilon(l)Edl \tag{6}$$

where $\varepsilon(l)$ is the strain, E is the elasticity modulus, and g_{33} is the piezoelectric voltage constant, given by $g_{33} = \frac{d_{33}}{\varepsilon_0 \varepsilon_r}$ with d_{33} being the piezoelectric coefficient, ε_r being the dielectric constant, and ε_0 being the permittivity of free space [15]. Hence, a higher piezoelectric coefficient and elastic modulus, and a lower dielectric constant result in higher nanogenerator performance. In this section, we will discuss the effect of fillers on composite-based nanogenerators. By varying the concentration, size, and morphology of fillers, the factors of composites can be controlled.

5.1 Piezoelectric coefficient

A piezoelectric coefficient is the most critical factor affecting the performance of nanogenerators. The piezoelectric coefficient (d_{33}) of bulk PVDF and its copolymers is remarkably low compared to that of bulk ceramics such as PZT (700 pC/N) [51], $BaTiO_3$ (350 pC/N) [52], and $Pb(Zn_{1/3}Nb_{2/3})O_3$-$PbTiO_3$ (PZN-PT) (2500 pC/N) [53]. As already mentioned in Section 2, the piezoelectric coefficient of PVDF and its copolymers is related to their crystallinity. In other words, the performance of nanogenerators can be improved by increasing the crystallinity of the ferroelectrics.

Recent studies have revealed that the crystallinity of ferroelectric polymers can change with fillers including organics, inorganics, and metals [16, 23, 39, 54]. The amount of β-phase in ferroelectric polymers can be enhanced by imbedding fillers into the ferroelectric polymer. For example, B. Dutta *et al.* reported a piezoelectric nanogenerator based on a NiO@SiO_2/PVDF nanocomposite [23]. To prevent agglomeration, they coated nonconductive SiO_2 on the NiO nanoparticles, which resulted in the homogeneous dispersion of the nanofillers in the polymer matrix. FTIR results indicated that the β-phase in the PVDF increased with the increasing loading fraction of NiO@SiO_2 as shown in **Figure 4a**. This phenomenon occurred due to the electrostatic interaction between the negatively charged surface of the NiO@SiO_2 nanoparticles and the positive $-CH_2$ groups of PVDF.

In addition to considering the content of fillers, the geometric structures of fillers should be also considered for their effect on the extent of β-phase crystallization and their influence on piezoelectric performance in the composites. A recent example was demonstrated in Ag/PVDF composites, where the effect of mixing silver fillers with different morphologies into P(VDF-TrFE) was investigated [54]. When Ag nanoparticles were added to P(VDF-TrFE), the crystalline transformation of P(VDF-TrFE) was not observed (**Figure 4b**). Similarly, work conducted by H. Paik *et al.* also found that the effect of adding Ag nanoparticles has an insignificant effect on the internal structures of the film [55]. On the other hand, when Ag nanowires with a diameter of 120 nm were mixed with P(VDF-TrFE), some α-phase transformed into β-phase, which indicates that the piezoelectric response can be enhanced as shown in **Figure 4c**. This was attributed to the fact that the molecular structure of P(VDF-TrFE) can be easily packed in a form that favors the formation of β-phase due to the planar surface of the Ag nanowires.

Interestingly, many research groups have used ferroelectric materials as fillers to enhance the performance of their nanogenerators. Even though ferroelectric fillers

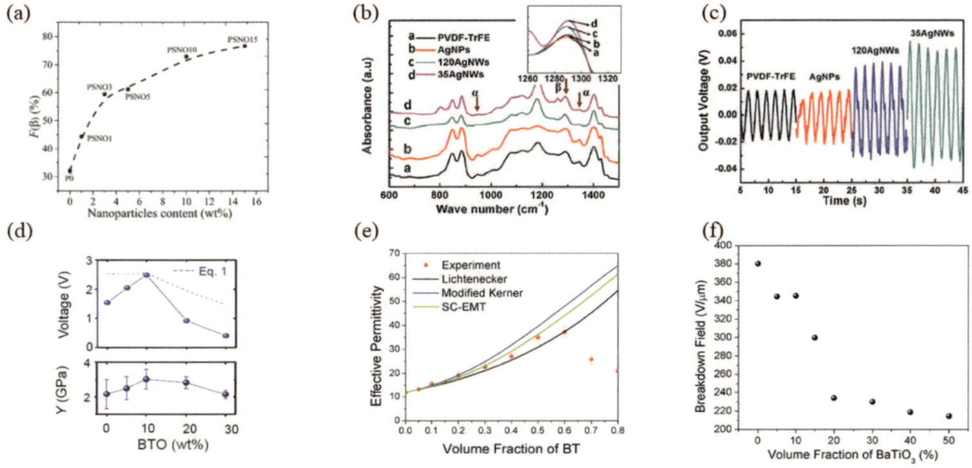

Figure 4.
(a) Variation of β-phase fraction (F(β)) with NiO@SiO$_2$ nanoparticles loading in PVDF matrix. Reproduced with permission [23]. Copyright 2018, American Chemical Society, (b) FTIR spectra of P(VDF-TrFE) film blended with different Ag nanocomposites. (the small picture was zoom-in view of peak intensity at 1290 cm^{-1}). (c) the piezoelectric response of P(VDF-TrFE) with different Ag nanofillers. Reproduced with permission [54]. Copyright 2016, Elsevier. (d) Comparison of the piezoelectric outputs with measured Young's modulus. Reproduced with permission [58]. Copyright 2018, MDPI. (e) Comparison of measured effective relative permittivity (at 1 kHz) of nanocomposites as a function of PFBPABT nanoparticle volume fraction with predicted values from different theoretical models. (f) the breakdown strengths (failure probabilities: 63.2%) at each volume fraction as determined from the Weibull analysis. Reproduced with permission [60]. Copyright 2018, American Chemical Society.

such as particles, nanowires, and nanorods have a poor piezoelectric coefficients, compared to that of bulk ceramics, they are still good candidates for piezoelectric nanogenerators. For example, PMN-PT nanofiber has a piezoelectric coefficient (d_{33}) of 50 pC/N [56].

However, PVDF and its copolymers have a negative piezoelectric coefficient, as previously mentioned. If PVDF and its copolymers are used as matrices, the conflict of piezoelectric constants can limit its advantages and result in lower output. Previous reports indicated that PZT/P(VDF-TrFE) 0–3 composites with 50 volume percent ceramic did not exhibit a piezoelectric response because they had the opposite sign piezoelectric coefficient, thus canceled out [57]. To circumvent this effect, different poling process can be adapted. For example, C. K. Jeong *et al*. demonstrated an enhanced nanogenerator based on $BaTiO_3$ nanowires and P(VDF-TrFE), which was poled using double-side poling process to remove the piezoelectric discrepancy between $BaTiO_3$ and P(VDF-TrFE) [21]. These examples further highlight why the mechanism behind the combination of piezoelectric nanofillers and piezoelectric polymers should be comprehensively explored.

5.2 Elastic modulus

It is also important to consider elastic modulus, the resistance to elastic deformation under load, according to the equation (Eq. (6)). A recent publication showed that Young's modulus plays an important role in a piezoelectric nanogenerator [58]. In that work, H. S. Kim *et al*. used paraelectric $BaTiO_3$ nanoparticles as fillers to exclude the effect from the piezoelectricity of the fillers. The Young's modulus increased from 2.17 to 3.03 GPa when the BTO content increased from 0 to 10 wt% and then decreased to 2.15 GPa as the content further increased to 30 wt% (**Figure 4d**). They found that the BTO content-output relationship was directly related to Young's modulus. Clearly, the optimal content for achieving the maximum Young's modulus, which corresponds to the highest piezoelectric output, should be identified.

5.3 Dielectric constant

Dielectric constants are an important parameter that also can affect the performance of nanogenerators. In composites, the dielectric constant can change with the volume of fillers. In a $BaTiO_3$-PVDF composite film, the dielectric constant increased from 8 to 31.8 at a frequency of 10^4 kHz when the volume fraction of fillers was increased from 0–30% [59]. Furthermore, the size of the nanoparticles can modify the dielectric properties. When the BaTiO3 filler size was increased from 10.5 to 34.6 nm, the dielectric constant at a frequency of 10^4 kHz increased from 20.1 to 31. In another study, P. Kim *et al.* systematically studied the dielectric constant and dielectric breakdown strength of a composite comprising P(VDF-HFP) and phosphoric acid surface-modified $BaTiO_3$ nanoparticles [60]. The dielectric constant increased as the nanoparticle volume fraction was increased from 0–60% and then decreased as the nanoparticles volume fraction increased above 60% (**Figure 4e**). The drop in dielectric constant at high volume fractions was attributed to the effect of the porosity of the composites. The dielectric breakdown strength decreased when the nanoparticle volume fraction increased, due to percolation of the $BaTiO_3$ nanoparticles and air voids (**Figure 4f**).

6. Energy sources

In addition to mechanical deformation, such as uniaxial compression and bending, energy sources such as air flow, sound, and thermal fluctuation are also available. A typical example of ambient natural sources is wind. Ferroelectric polymer-based nanogenerators can easily yield electricity from wind energy because of their considerable flexibility. For example, Li *et al.* fabricated nanogenerators based on P(VDF-TrFE) films to harvest mechanical energy from low speed wind, which is very close to winds commonly observed in the natural environment [34]. The experimental setups are illustrated in **Figure 5a**. They evaluated the wind energy extraction capability by comparing three instinct operating modes. Each operating modes exhibited different piezoelectric energy harvesting behaviors, including mode frequency, output voltage, and power. The authors concluded that a nanogenerator operating where the wind is incident to the side of the film delivered high output power compared to other modes within the wind speed range of 1.7 and 4.7 m/s (**Figure 5b**). These results may be attributed to the huge deflection that occurred over the entire area of the nanogenerator.

Besides wind, sound is also an abundant energy source that can be found in the environment. PVDF films can be used as an active layer for harvesting sound waves. A variety of polymer-based nanogenerators have been developed to scavenge energy from sonic inputs. For example, S. Cha *et al.* presented nanoporous arrays of PVDF fabricated by a lithography-free, template-assisted preparation method (**Figure 5c**) [61]. They demonstrated that the porous PVDF nanogenerators generated an output voltage and current of 2.6 V and 0.6 μA, respectively. These output values were 5.2 times and 6 times higher than bulk PVDF film nanogenerators, respectively (**Figure 5d**). The piezoelectric potential was enhanced in the nanoporous structure as a result of the geometrical strain confinement effect.

More recently, nanoweb-type nanogenerators were developed to convert sound wave into electrical energy [62, 63]. The benefit of using nanowebs in nanogenerators compared to films is their high power generation because they are more flexible and easier to vibrate under the same acoustic waves.

Pyroelectricity is a property of certain materials, which develop an electric field across the polar axis when there is a temperature change. According to the pyroelectric theory [64], the pyroelectric current I is expressed as

$$I = \frac{dQ}{dt} = S\lambda\frac{dT}{dt} \tag{7}$$

where Q, S, λ, and T are the induced charge, the electrode surface of the material sample, the pyroelectric coefficient, and the temperature, respectively. Ferroelectric ceramics like PMN-PT and PZT have high pyroelectric coefficients above 550 $\mu Cm^{-2} K^{-1}$ compared to those of ferroelectric polymers ($33 \mu Cm^{-2} K^{-1}$). As a result, a PZT-based nanogenerator delivered more current density than that of ferroelectric polymers. However, PVDF and its copolymers have also been employed to harvest thermal energy from temperature fluctuations, where they provide the benefits of light weight, low-cost, lead-free, flexible, and transparent

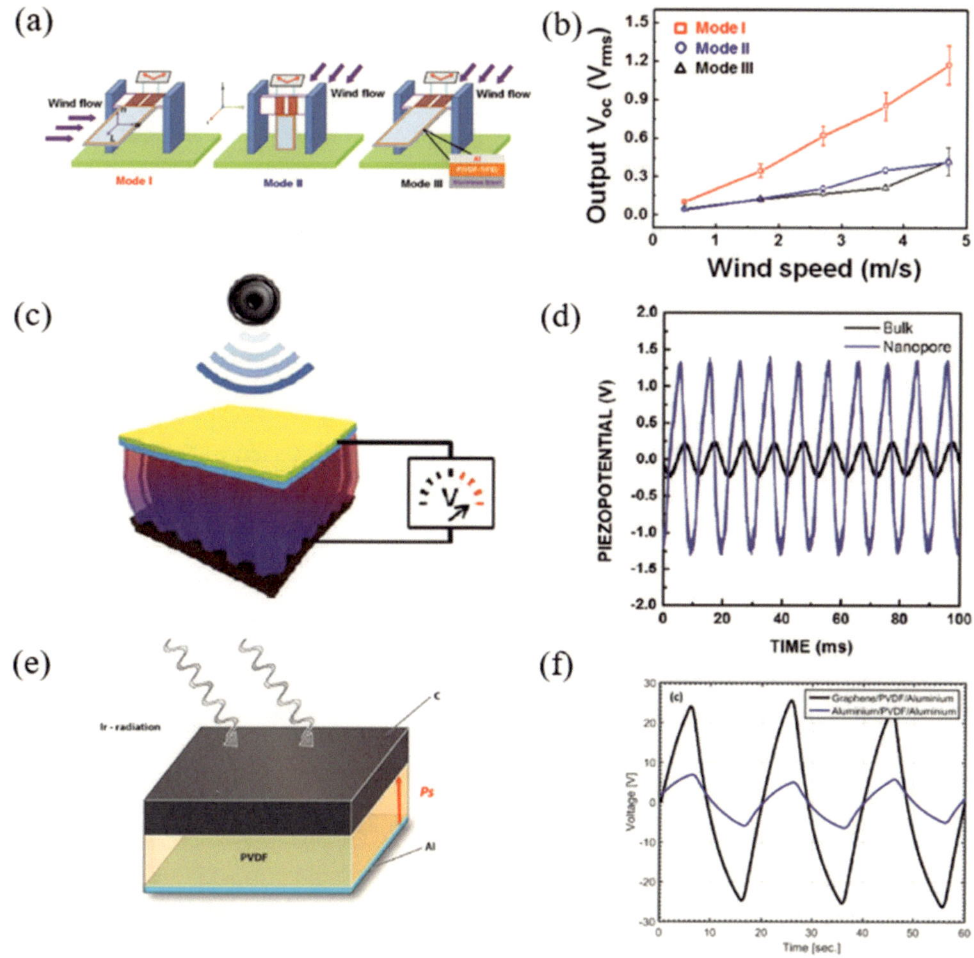

Figure 5.
(a) Schematic illustration of three different operation modes: Mode I, mode II, and mode III. (b) Output voltage (Vout) as a function of incident wind speed within the range of 0.5–4.7 m/s for modes I, II, and III. Reproduced with permission [34]. Copyright 2014, AIP publishing. (c) Schematic depiction of the potential of porous nanogenerators generated by the sonic wave. (d) Piezoelectric potential from the porous PVDF and bulk structure under the same force. Reproduced with permission [61]. Copyright 2011, American Chemical Society. (e) Schematic illustration of a PVDF/graphene ink pyroelectric device. (f) Open-circuit voltage for aluminum electrode and PVDF/graphene ink pyroelectric energy harvesting device. Reproduced with permission [68]. Copyright 2017, American Chemical Society.

properties [41, 65, 66]. A number of research groups have focused on improving the performance of such nanogenerators by changing their designed conditions. For example, D. Zabek *et al*. reported a simple modification of the electrode design in a PVDF-based nanogenerator [67]. The partially covered electrode was developed by photo-lithography and a wet etching process. The modification of the electrode design enabled a significant increase in pyroelectric voltage (380%), current (480%), and pyroelectric energy harvesting transformation (1080%).Their findings showed that the electrode area coverage could be optimized by balancing the ratio of the area of aluminum electrode required to get the free charge to the area of exposed PVDF to improve heat transfer. In another study, they also demonstrated that interconnected graphene nanoplates (GNPs) can be used as an electrode for high thermal radiation absorbance and high electrical conductivity as shown in **Figure 5e** [68]. The GNPs were easily deposited by a screen print technique. The graphene ink/PVDF/Al system showed better performance than a system with conventional aluminum electrodes (**Figure 5f**).

7. Biocompatibility

Lead (Pb) is considered one of the core materials in modern society because it is inexpensive and has high density and resistance to corrosion. However, it is also toxic and can harm multiple human body systems. Legislation has been adopted in many countries to prevent and reduce the use of Pb in many communities [69]. Currently, many of the materials with high piezoelectric performance are lead-based materials, that is, $PbTiO_3$, $Pb(Zr,Ti)O_3$, PMN-PT, $Pb(Mg_{1/3}Nb_{2/3})O3\text{-}Pb(Zr,Ti)O_3$ (PMN-PZT), and PIN-PMN-PT [70–72]. They are used in a wide variety of applications including sensors, actuators, and ultrasonic transducers. The EU's Restriction of Hazardous Substances (RoHS) Directive of 2002 and its revision (RoHS 2) in 2011 designated certain piezoelectric devices as exemptions since lead-free piezoelectrics cannot completely replace all lead-based piezoelectrics at this time [69]. Nevertheless, it is urgent to develop lead-free piezoelectrics that can perform as well as or better than PZT.

From the biocompatibility point of view, one of the most promising alternatives is a ferroelectric polymer. The unique advantage of PVDF and its copolymers lies in their good stability, similar to PTFE. They are chemically inert and resistant to sunlight [65, 73, 74]. There have been reports on the biocompatibility of PVDF-based nanogenerators. For example, Y. Yu *et al*. demonstrated biomechanical energy harvesting in a living organism using PVDF-based nanogenerator, where a

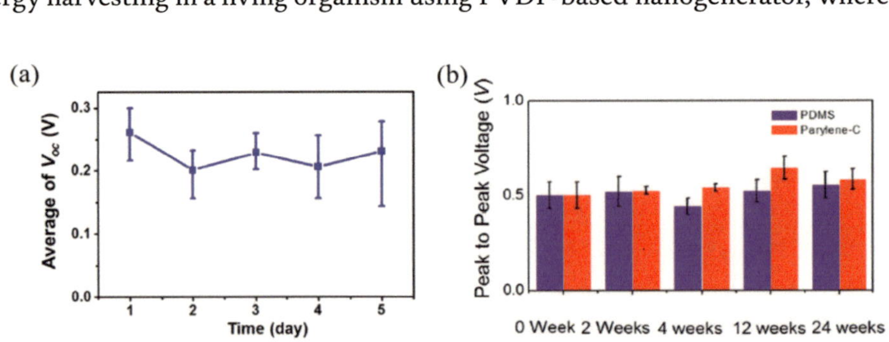

Figure 6.
(a) The variation of voltage output as a function of the testing period. Reproduced with permission [75]. Copyright 2016, Elsevier. (b) Stability of voltage outputs as a function of time for implantable NGs with different packaging strategies. The purple and red bars represent implantable NGs with PDMS and PDMS/Parylene-C package, respectively. Reproduced with permission [76]. Copyright 2018, Elsevier.

mesoporous PVDF film was encapsulated by polydimethylsiloxane (PDMS) [75]. They inserted the nanogenerator under the skin of a mice's leg and observed that there were no signs of incompatibility induced by the NG. In addition, an output voltage of 200 mV was stably generated from the gentle motions of the rat's leg for 5 days as shown in **Figure 6a**. In another work, J. Li *et al*. studied PDMS and PDMS/Parylene-C packaged PVDF nanogenerators implanted inside a female ICR mouse for up to 6 months [76]. They observed no signs of toxicity or incompatibility via pathological analyses and blood and serum test. In addition, there was no issue with reduction in electrical outputs over the entire implantation period (**Figure 6b**). Therefore, it is conceivable that PVDF-based nanogenerators can be used in a biological environment.

8. Conclusions

In this chapter, we summarized recent studies of ferroelectric polymer PVDF-based nanogenerators. PVDF and its copolymers are attractive materials for nanogenerator applications. The materials are flexible, transparent, chemically stable, easy to process, and biocompatible. Because of these advantages, PVDF-based nanogenerators can be placed anywhere, including bones, human skins, and wearable devices that usually have curved surfaces. In addition, they can harvest electricity from a variety of energy sources.

Considerable effort has been expended by numerous research groups to enhance the performance of PVDF-based nanogenerators using various fabrication methods, designing device structures, and incorporating fillers. However, studies on PVDF-based nanogenerators are fairly limited. The mechanisms leading to the enhanced performance after the incorporation of fillers should be comprehensively surveyed. Especially, in the case of ferroelectric fillers, a careful approach to the enhancement should be taken, since a piezoelectric effect also occurs with fillers. To optimize nanogenerator performance, circuits for nanogenerators should be developed. Their performance depends on the circuits and components, including load resistors, capacitors, and wiring. In order to realize the commercialization of PVDF-based nanogenerators, packaging of the devices is critical to prevent mechanical fatigue. The selection of substrates and electrode materials can also improve the mechanical properties of these devices. Therefore, the relationship between a ferroelectric polymer layer and other layers should be considered.

We hope that this chapter will help readers to better understand principles of PVDF-based nanogenerators. Furthermore, several new approaches that have been in this chapter can be adopted for other applications, such as sensors, actuators, and field-effect transistors.

Acknowledgements

This chapter is based on a research that has been conducted as part of the KAIST funded Global Singularity Research Program for 2019. Dr. Panpan Li was also supported by the Korea Research Fellowship Program funded by the National Research Foundation of Korea (no. 2017H1D3A1A01054478).

Author details

Jeongjae Ryu[1], Seongmun Eom[1], Panpan Li[1], Chi Hao Liow[1] and Seungbum Hong[1,2]*

1 Department of Materials Science and Engineering, Korea Advanced Institute of Science and Technology (KAIST), Daejeon, Republic of Korea

2 KAIST Institute of the NanoCentury, Korea Advanced Institute of Science and Technology, Daejeon, Republic of Korea

*Address all correspondence to: seungbum@kaist.ac.kr

References

[1] Ang L-M, Seng KP. Big sensor data applications in urban environments. Big Data Research. 2016;**4**:1-12. DOI: 10.1016/j.bdr.2015.12.003

[2] Palanisamy V, Thirunavukarasu R. Implications of big data analytics in developing healthcare frameworks—A review. Journal of King Saud University - Computer and Information Sciences. 2017;**31**(4):415-425. DOI: 10.1016/j.jksuci.2017.12.007

[3] Risteska Stojkoska BL, Trivodaliev KV. A review of internet of things for smart home: Challenges and solutions. Journal of Cleaner Production. 2017;**140**:1454-1464. DOI: 10.1016/j.jclepro.2016.10.006

[4] Attal F, Mohammed S, Dedabrishvili M, Chamroukhi F, Oukhellou L, Amirat Y. Physical human activity recognition using wearable sensors. Sensors (Basel). 2015;**15**:31314-31338. DOI: 10.3390/s151229858

[5] Hao Y, Foster R. Wireless body sensor networks for health-monitoring applications. Physiological Measurement. 2008;**29**:R27-R56. DOI: 10.1088/0967-3334/29/11/R01

[6] Lee Y-D, Chung W-Y. Wireless sensor network based wearable smart shirt for ubiquitous health and activity monitoring. Sensors and Actuators B: Chemical. 2009;**140**:390-395. DOI: 10.1016/j.snb.2009.04.040

[7] Yamada T, Hayamizu Y, Yamamoto Y, Yomogida Y, Izadi-Najafabadi A, Futaba DN, et al. Stretchable nanotube strain sensor for human-motion detection. Nature Nanotechnology. 2011;**6**:296-301. DOI: 10.1038/nnano.2011.36

[8] Kim D, Hong S, Hong J, Choi Y-Y, Kim J, Park M, et al. Fabrication of vertically aligned ferroelectric Polyvinylidene fluoride Mesoscale rod arrays. Journal of Applied Polymer Science. 2013;**130**:3842-3848. DOI: 10.1002/app.39415

[9] Oh S, Kim Y, Choi YY, Kim D, Choi H. No K, Fabrication of vertically well-aligned P(VDF-TrFE) nanorod arrays. Advanced Materials. 2012;**24**:5708-5712. DOI: 10.1002/adma.201201940

[10] Bune AV, Fridkin VM, Ducharme S, Blinov LM, Palto SP, Sorokin AV, et al. Two-dimensional ferroelectric films. Nature. 1998;**391**:874-877. DOI: 10.1038/36069

[11] Hu Z, Tian M, Nysten B, Jonas AM. Regular arrays of highly ordered ferroelectric polymer nanostructures for non-volatile low-voltage memories. Nature Materials. 2009;**8**:62-67. DOI: 10.1038/nmat2339

[12] Pi Z, Zhang J, Wen C, Zhang Z-B, Wu D. Flexible piezoelectric nanogenerator made of poly(vinylidenefluoride-co-trifluoroethylene) (PVDF-TrFE) thin film. Nano Energy. 2014;**7**:33-41. DOI: 10.1016/j.nanoen.2014.04.016

[13] Jung W-S, Lee M-J, Kang M-G, Moon HG, Yoon S-J, Baek S-H, et al. Powerful curved piezoelectric generator for wearable applications. Nano Energy. 2015;**13**:174-181. DOI: 10.1016/j.nanoen.2015.01.051

[14] Soin N, Shah TH, Anand SC, Geng J, Pornwannachai W, Mandal P, et al. Novel "3-D spacer" all fibre piezoelectric textiles for energy harvesting applications. Energy & Environmental Science. 2014;**7**:1670-1679. DOI: 10.1039/c3ee43987a

[15] Alluri NR, Saravanakumar B, Kim SJ. Flexible, hybrid piezoelectric film ($BaTi_{(1-x)}Zr_{(x)}O_3$)/PVDF

nanogenerator as a self-powered fluid velocity sensor. ACS Applied Materials & Interfaces. 2015;7:9831-9840. DOI: 10.1021/acsami.5b01760

[16] Kar E, Bose N, Dutta B, Banerjee S, Mukherjee N, Mukherjee S. 2D SnO_2 nanosheet/PVDF composite based flexible, self-cleaning piezoelectric energy harvester. Energy Conversion and Management. 2019;**184**:600-608. DOI: 10.1016/j.enconman.2019.01.073

[17] Thakur P, Kool A, Hoque NA, Bagchi B, Khatun F, Biswas P, et al. Superior performances of In situ synthesized ZnO/PVDF thin film based self-poled piezoelectric nanogenerator and self-charged photo-power Bank with high durability. Nano Energy. 2018;**44**:456-467. DOI: 10.1016/j.nanoen.2017.11.065

[18] Saravanakumar B, Soyoon S, Kim SJ. Self-powered pH sensor based on a flexible organic-inorganic hybrid composite nanogenerator. ACS Applied Materials & Interfaces. 2014;**6**:13716-13723. DOI: 10.1021/am5031648

[19] Karan SK, Bera R, Paria S, Das AK, Maiti S, Maitra A, et al. An approach to design highly durable piezoelectric nanogenerator based on self-poled PVDF/AlO-rGO flexible Nanocomposite with high power density and energy conversion efficiency. Advanced Energy Materials. 2016;**6**:1601016. DOI: 10.1002/aenm.201601016

[20] Yaqoob U, Uddin ASMI, Chung G-S. A novel tri-layer flexible piezoelectric nanogenerator based on surface-modified graphene and PVDF-$BaTiO_3$ nanocomposites. Applied Surface Science. 2017;**405**:420-426. DOI: 10.1016/j.apsusc.2017.01.314

[21] Jeong CK, Baek C, Kingon AI, Park KI, Kim SH. Lead-free perovskite nanowire-employed piezopolymer for highly efficient flexible nanocomposite energy harvester. Small. 2018;**14**:e1704022. DOI: 10.1002/smll.201704022

[22] Zhang C, Fan Y, Li H, Li Y, Zhang L, Cao S, et al. Fully rollable lead-free poly(vinylidene fluoride)-niobate-based nanogenerator with ultra-flexible nano-network electrodes. ACS Nano. 2018;**12**:4803-4811. DOI: 10.1021/acsnano.8b01534

[23] Dutta B, Kar E, Bose N, Mukherjee S.NiO@SiO_2/PVDF: A flexible polymer nanocomposite for a high performance human body motion-based energy harvester and tactile e-skin Mechanosensor. ACS Sustainable Chemistry & Engineering. 2018;**6**:10505-10516. DOI: 10.1021/acssuschemeng.8b01851

[24] Maity K, Mandal D. All-organic high-performance piezoelectric nanogenerator with multilayer assembled electrospun nanofiber mats for self-powered multifunctional sensors. ACS Applied Materials & Interfaces. 2018;**10**:18257-18269. DOI: 10.1021/acsami.8b01862

[25] Li J, Chen S, Liu W, Fu R, Tu S, Zhao Y, et al. High performance piezoelectric nanogenerators based on electrospun ZnO Nanorods/poly(vinylidene fluoride) composite membranes. The Journal of Physical Chemistry C. 2019;**123**:11378-11387. DOI: 10.1021/acs.jpcc.8b12410

[26] Cui Z, Hassankiadeh NT, Zhuang Y, Drioli E, Lee YM. Crystalline polymorphism in poly (vinylidenefluoride) membranes. Progress in Polymer Science. 2015;**51**:94-126. DOI: 10.1016/j.progpolymsci.2015.07.007

[27] Li M, Wondergem HJ, Spijkman MJ, Asadi K, Katsouras I, Blom PW, et al. Revisiting the delta-phase of

poly(vinylidene fluoride) for solution-processed ferroelectric thin films. Nature Materials. 2013;**12**: 433-438. DOI: 10.1038/nmat3577

[28] He X, Yao K, Gan BK. Phase transition and properties of a ferroelectric poly(vinylidene fluoride-hexafluoropropylene) copolymer. Journal of Applied Physics. 2005;**97**:084101. DOI: 10.1063/1.1862323

[29] Wegener M, Künstler W, Richter K, Gerhard-Multhaupt R. Ferroelectric polarization in stretched Piezo- and pyroelectric poly(vinylidene fluoride-hexafluoropropylene) copolymer films. Journal of Applied Physics. 2002;**92**:7442-7447. DOI: 10.1063/1.1524313

[30] Jayasuriya AC, Scheinbeim JI. Ferroelectric behavior in solvent cast poly(vinylidene fluoridehexafluoropropylene) copolymer films. Applied Surface Science. 2001;**175-176**:386-390

[31] Kawai H. The piezoelectricity of poly(vinylidene fluoride). Japanese Journal of Applied Physics. 1969;**8**:975-976

[32] Katsouras I, Asadi K, Li M, van Driel TB, Kjaer KS, Zhao D, et al. The negative piezoelectric effect of the ferroelectric polymer poly(vinylidene fluoride). Nature Materials. 2016;**15**:78-84. DOI: 10.1038/nmat4423

[33] Cai X, Lei T, Sun D, Lin L. A critical analysis of the α, β and γ phases in poly(vinylidene fluoride) using FTIR. RSC Advances. 2017;**7**:15382-15389. DOI: 10.1039/c7ra01267e

[34] Jun Li D, Hong S, Gu S, Choi Y, Nakhmanson S, Heinonen O, et al. Polymer piezoelectric energy harvesters for low wind speed. Applied Physics Letters. 2014;**104**:012902. DOI: 10.1063/1.4861187

[35] Kim D, Hong S, Li D, Roh HS, Ahn G, Kim J, et al. A spring-type piezoelectric energy harvester. RSC Advances. 2013;**3**:3194-3198. DOI: 10.1039/c2ra22554a

[36] Ryu J, Kim J, Oh J, Lim S, Sim JY, Jeon JS, et al. Intrinsically stretchable multi-functional Fiber with energy harvesting and strain sensing capability. Nano Energy. 2019;**55**:348-353. DOI: 10.1016/j.nanoen.2018.10.071

[37] Mandal D, Yoon S, Kim KJ. Origin of piezoelectricity in an electrospun poly(vinylidene fluoride-trifluoroethylene) nanofiber web-based nanogenerator and nano-pressure sensor. Macromolecular Rapid Communications. 2011;**32**:831-837. DOI: 10.1002/marc.201100040

[38] Egusa S, Wang Z, Chocat N, Ruff ZM, Stolyarov AM, Shemuly D, et al. Multimaterial piezoelectric fibres. Nature Materials. 2010;**9**:643-648. DOI: 10.1038/nmat2792

[39] Mao Y, Zhao P, McConohy G, Yang H, Tong Y, Wang X. Sponge-like piezoelectric polymer films for scalable and integratable nanogenerators and self-powered electronic systems. Advanced Energy Materials. 2014;**4**:1301624. DOI: 10.1002/aenm.201301624

[40] Madaeni SS, Taheri AH. Effect of casting solution on morphology and performance of PVDF microfiltration membranes. Chemical Engineering & Technology. 2011;**34**:1328-1334. DOI: 10.1002/ceat.201000177

[41] Lee JH, Lee KY, Gupta MK, Kim TY, Lee DY, Oh J, et al. Highly stretchable piezoelectric-pyroelectric hybrid nanogenerator. Advanced Materials. 2014;**26**:765-769. DOI: 10.1002/adma.201303570

[42] Bhavanasi V, Kusuma DY, Lee PS. Polarization orientation,

piezoelectricity, and energy harvesting performance of ferroelectric PVDF-TrFE nanotubes synthesized by nanoconfinement. Advanced Energy Materials. 2014;**4**:1400723. DOI: 10.1002/aenm.201400723

[43] Calahorra Y, Whiter RA, Jing Q, Narayan V, Kar-Narayan S. Localized electromechanical interactions in ferroelectric P(VDF-TrFE) nanowires investigated by scanning probe microscopy. APL Materials. 2016;**4**:116106. DOI: 10.1063/1.4967752

[44] Chen X, Shao J, An N, Li X, Tian H, Xu C, et al. Self-powered flexible pressure sensors with vertically well-aligned piezoelectric nanowire arrays for monitoring vital signs. Journal of Materials Chemistry C. 2015;**3**:11806-11814. DOI: 10.1039/c5tc02173a

[45] Xin Y, Zhu J, Sun H, Xu Y, Liu T, Qian C. A brief review on piezoelectric PVDF nanofibers prepared by electrospinning. Ferroelectrics. 2018;**526**:140-151. DOI: 10.1080/00150193.2018.1456304

[46] Chang C, Limkrailassiri K, Lin L. Continuous near-field electrospinning for large area deposition of orderly nanofiber patterns. Applied Physics Letters. 2008;**93**:123111. DOI: 10.1063/1.2975834

[47] Chang C, Tran VH, Wang J, Fuh Y-K, Lin L. Direct-write piezoelectric polymeric nanogenerator with high energy conversion efficiency. Nano Letters. 2010;**10**:726-731. DOI: 10.1021/nl9040719

[48] Lestoquoy G, Chocat N, Wang Z, Joannopoulos JD, Fink Y. Fabrication and characterization of thermally drawn fiber capacitors. Applied Physics Letters. 2013;**102**:152908. DOI: 10.1063/1.4802783

[49] Lu X, Qu H, Skorobogatiy M. Piezoelectric micro- and nanostructured fibers fabricated from thermoplastic nanocomposites using a fiber drawing technique: Comparative study and potential applications. ACS Nano. 2017;**11**:2103-2114. DOI: 10.1021/acsnano.6b08290

[50] Zhao J, You Z, Shoe A. Embedded piezoelectric energy harvester for wearable sensors. Sensors. 2014;**14**:12497-12510. DOI: 10.3390/s140712497

[51] Jaffe H, Berlincourt DA. Piezoelectric transducer materials. Proceedings of the IEEE. 1965;**53**:1372-1386. DOI: 10.1109/PROC.1965.4253

[52] Takahashi H, Numamoto Y, Tani J, Matsuta K, Qiu J, Tsurekawa S. Lead-free barium titanate ceramics with large piezoelectric constant fabricated by microwave sintering. Japanese Journal of Applied Physics. 2006;**45**:L30-L32. DOI: 10.1143/jjap.45.L30

[53] Park S-E, Shrout TR. Ultrahigh strain and piezoelectric behavior in relaxor based ferroelectric single crystals. Journal of Applied Physics. 1997;**82**:1804-1811. DOI: 10.1063/1.365983

[54] Chen H-J, Han S, Liu C, Luo Z, Shieh H-PD, Hsiao R-S, et al. Investigation of PVDF-TrFE composite with nanofillers for sensitivity improvement. Sensors and Actuators A: Physical. 2016;**245**:135-139. DOI: 10.1016/j.sna.2016.04.056

[55] Paik H, Choi YY, Hong S, No K. Effect of Ag nanoparticle concentration on the electrical and ferroelectric properties of Ag/P(VDF-TrFE) composite films. Scientific Reports. 2015;**5**:13209. DOI: 10.1038/srep13209

[56] Xu S, Poirier G, Yao N. Fabrication and piezoelectric property of PMN-PT nanofibers. Nano Energy. 2012;**1**:602-607. DOI: 10.1016/j.nanoen.2012.03.011

[57] Chan HLW, Chen Y, Choy CL. A poling study of PZT/P(VDF-TrFE) copolymer 0-3 composites. Integrated Ferroelectrics. 2006;**9**:207-214. DOI: 10.1080/10584589508012925

[58] Kim HS, Lee DW, Kim DH, Kong DS, Choi J, Lee M, et al. Dominant role of Young's modulus for electric power generation in PVDF-$BaTiO_3$ composite-based piezoelectric nanogenerator. Nanomaterials (Basel). 2018;**8**:777. DOI: 10.3390/nano8100777

[59] Kobayashi Y, Tanase T, Tabata T, Miwa T, Konno M. Fabrication and dielectric properties of the $BaTiO_3$–polymer nano-composite thin films. Journal of the European Ceramic Society. 2008;**28**:117-122. DOI: 10.1016/j.jeurceramsoc.2007.05.007

[60] Kim P, Doss NM, Tillotson JP, Hotchkiss PJ, Pan M-J, Marder SR, et al. High energy density nanocomposites based on surface-modified $BaTiO_3$ and a ferroelectric polymer. ACS Nano. 2009;**3**:2581-2592. DOI: 10.1021/nn9006412

[61] Cha S, Kim SM, Kim H, Ku J, Sohn JI, Park YJ, et al. Porous PVDF as effective sonic wave driven nanogenerators. Nano Letters. 2011;**11**:5142-5147. DOI: 10.1021/nl202208n

[62] Lang C, Fang J, Shao H, Ding X, Lin T. High-sensitivity acoustic sensors from nanofibre webs. Nature Communications. 2016;**7**:11108. DOI: 10.1038/ncomms11108

[63] Sun B, Li X, Zhao R, Ji H, Qiu J, Zhang N, et al. Electrospun poly(vinylidene fluoride)-zinc oxide hierarchical composite Fiber membrane as piezoelectric acoustoelectric nanogenerator. Journal of Materials Science. 2018;**54**:2754-2762. DOI: 10.1007/s10853-018-2985-x

[64] Cuadras A, Gasulla M, Ferrari V. Thermal energy harvesting through pyroelectricity. Sensors and Actuators A: Physical. 2010;**158**:132-139. DOI: 10.1016/j.sna.2009.12.018

[65] Vuorinen T, Zakrzewski M, Rajala S, Lupo D, Vanhala J, Palovuori K, et al. Printable, transparent, and flexible touch panels working in sunlight and moist environments. Advanced Functional Materials. 2014;**24**:6340-6347. DOI: 10.1002/adfm.201401140

[66] Talemi P, Delaigue M, Murphy P, Fabretto M. Flexible polymer-on-polymer architecture for piezo/pyroelectric energy harvesting. ACS Applied Materials & Interfaces. 2015;**7**:8465-8471. DOI: 10.1021/am5089082

[67] Zabek D, Taylor J, Boulbar EL, Bowen CR. Micropatterning of flexible and free standing polyvinylidene difluoride (PVDF) films for enhanced pyroelectric energy transformation. Advanced Energy Materials. 2015;**5**:1401891. DOI: 10.1002/aenm.201401891

[68] Zabek D, Seunarine K, Spacie C, Bowen C. Graphene ink laminate structures on poly(vinylidene difluoride) (PVDF) for pyroelectric thermal energy harvesting and waste heat recovery. ACS Applied Materials & Interfaces. 2017;**9**:9161-9167. DOI: 10.1021/acsami.6b16477

[69] Bell AJ, Deubzer O. Lead-free piezoelectrics—The environmental and regulatory issues. MRS Bulletin. 2018;**43**:581-587. DOI: 10.1557/mrs.2018.154

[70] Setter N, Damjanovic D, Eng L, Fox G, Gevorgian S, Hong S, et al. Ferroelectric thin films: Review of materials, properties, and applications. Journal of Applied Physics. 2006;**100**:051606. DOI: 10.1063/1.2336999

[71] Kim J, Hong S, Bühlmann S, Kim Y, Park M, Kim YK. et al. Effect

of deposition temperature of TiO_2 on the piezoelectric property of $PbTiO_3$ film grown by PbO gas phase reaction sputtering. Journal of Applied Physics. 2010;**107**:104112. DOI: 10.1063/1.3406148

[72] Panda PK, Sahoo B. PZT to lead free piezo ceramics: A review. Ferroelectrics. 2015;**474**:128-143. DOI: 10.1080/00150193.2015.997146

[73] Bayer IS, Tiwari MK, Megaridis CM. Biocompatible poly(vinylidene fluoride)/ cyanoacrylate composite coatings with tunable hydrophobicity and bonding strength. Applied Physics Letters. 2008;**93**:173902. DOI: 10.1063/1.3009292

[74] Guo HF, Li ZS, Dong SW, Chen WJ, Deng L, Wang YF, et al. PVDF electrospun scaffolds for wound healing applications. Colloids and Surfaces B: Biointerfaces. 2012;**96**:29-36. DOI: 10.1016/j.colsurfb.2012.03.014

[75] Yu Y, Sun H, Orbay H, Chen F, England CG, Cai W, et al. Biocompatibility and in vivo operation of implantable mesoporous PVDF-based nanogenerators. Nano Energy. 2016;**27**:275-281. DOI: 10.1016/j.nanoen.2016.07.015

[76] Li J, Kang L, Yu Y, Long Y, Jeffery JJ, Cai W, et al. Study of Long-term biocompatibility and bio-safety of implantable nanogenerators. Nano Energy. 2018;**51**:728-735. DOI: 10.1016/j.nanoen.2018.07.008

Chapter 5

Piezoelectricity in Self-Assembled Peptides: A New Way towards Electricity Generation at Nanoscale

Vladislav Slabov, Svitlana Kopyl, Marco P. Soares dos Santos and Andrei Kholkin

Abstract

Self-assembled nanostructured peptides are of great interest nowadays due to their biocompatibility and an array of outstanding functional properties. Among them, strong piezoelectricity combined with low dielectric constant is beneficial for high voltage and power generation at the nanoscale. This Chapter is an overview of the piezoelectric phenomena in self-assembled peptides including effects of the growth conditions, self-assembly, and measurement techniques on their functional response as well as the origin of strong piezoelectricity in this material. The current status of electrical energy harvesting in self-assembled peptides useful for biomedical applications along with the challenges and perspectives for using these piezoelectric biomaterials will be discussed. This Chapter is expected to provide a guidance towards future design and application of novel functional self-assembled materials based on nanostructured peptides.

Keywords: self-assembly, peptide, piezoelectricity, nanogenerator, energy harvesting

1. Introduction

One of the most important trends in modern microelectronics has been the dramatic reduction in size, decrease in power consumption and, finally, increase in the functionality. However, the batteries supplying the energy for the devices did not decrease significantly in size over the last years (power density increased but only several times, [1]). They represent currently a significant volume and weight as compared to the overall size and weight of the device itself. As such, advances in low power design of many electronic components opened up the possibility of using energy from the environment to power these microdevices. This increases their functionality as they can be completely autonomous and wearable and will not require replacing batteries for the entire lifetime of the device. One of the major mechanisms to achieve electric energy harvesting is piezoelectric effect, which converts mechanical vibrations into electricity, and pyroelectric energy generation, which uses temperature changes. The vibration sources are quite abundant and can be found, e.g., in the human motion, blood flow, low-frequency vibrations of buildings, motion of the cars, and acoustic noise. These can be directly applied to piezoelectric material to generate electric power. In fact, piezoelectric harvesting allows easy miniaturization because piezoelectric effect is scale-independent and can be even higher at the nanoscale [2].

The interest in the application of piezoelectrics and pyroelectrics for electrical energy harvesting has grown significantly last years. Most piezoelectric harvesting sources can produce power on the order of milliwatts but it is still small for large systems. However, the generated power is just enough for hand-held and implanted devices and miniaturized wearable electronics (e.g. sensor networks). Piezoelectric energy harvesting has been investigated since the late '90s and it is currently considered as an emerging technology. Miniaturization of the devices requires using nanoscale dimensions, so that electricity must be harvested timely and exactly where it can be consumed. For example, it can be used for biomedical purposes, the piezoelectric scaffolds for cell regeneration being one of such applications [3] and for instrumented orthopedic implants [4].

Pyroelectric materials are a subclass of piezoelectric crystals with a polar structure. All pyroelectric materials are also piezoelectric, although the *vice versa* does not occur. They exhibit a change in polarization with temperature change [5]. It is well known that there is a noticeable temperature difference between the human body and the ambient environment (the average human body temperature being 37°C). Therefore, there is always a heat loss from the human body to the ambient environment, mainly through skin and through the process of respiration. This waste heat can be converted into electric energy using pyroelectric effect. Several pyroelectric nanogenerators have already been reported, which utilized pyroelectric/ferroelectric materials [6–8].

Inorganic ferroelectrics, such as lead zirconate titanate (PZT), lithium niobate (LNO) and barium titanate (BTO), are widely used as efficient piezoelectrics, pyroelectrics, memory cells, and electrooptic modulators [9]. They possess high switchable polarization, strong piezoelectric response, remarkable pyroelectric and electrooptic properties, but have some disadvantages, such as poor biocompatibility and brittle nature that limits their use in biomedical applications. Moreover, their processing requires high temperatures, so their miniaturization and integration with microelectromechanical systems (MEMS) is difficult. Novel organic or polymer materials (e.g. polyvinylidene fluoride, PVDF) seem to not have these disadvantages, but still may require a protection from the tissue or cell in the in the direct contact [10] because they are not biocompatible.

In this context, bioinspired materials with biocompatibility, easy preparation, nontoxicity and environmental friendliness have been regarded as promising alternatives [11]. Until recently, properties of organic and biomolecular piezoelectrics were considered by far inferior to those of their inorganic counterparts, thus making their applications as piezoelectrics and pyroelectrics unfeasible. Recent studies revealed that the softness of hydrogen bonds in some representative classes of biomolecular polar materials may be the origin of strong piezoelectric effects at room temperature [12]. It has been reported that the piezoelectric coefficients in self-assembled short peptides (diphenylalanine, NH_2-Phe-Phe-COOH) are practically the same as in popular transducer material LNO [13]. This discovery has led to a surge in research on piezoelectric biomaterials. A variety of microstructures and phases have been produced that could lead to the enhancement of physical and chemical properties by designing molecular structures, fabricating heterostructures and introducing dopants [14]. Several reports show that the piezoelectric response in biomaterials is directly related to their phase, shape and growth direction [15]. It has been found that piezoelectricity of peptide fibrils and phages in axial direction is stronger than that in the radial direction [15]. Electric charges generated by the physical stimulation of piezoelectric biomaterials were found to stimulate bone growth, wound healing and tissue regeneration [16].

This Chapter is an overview of the piezoelectric phenomena in self-assembled peptides, focusing on diphenylalanine, and includes effects of the growth conditions, self-assembly, and polarization of self-assembled nanostructures peptides on their piezoelectric properties. The discussion will further focus on their applications as nanogenerators and energy-storage devices.

2. Introduction to self-assembled peptides

Supramolecular self-assembly plays a significant role in building highly ordered and functional structures in biology. Natural biological tissues are hierarchically structured, and these structures appear to correlate strongly with tissue properties and functionalities. A single macromolecule can form various functional structures when self-assembled under different conditions. Nanotechnology is trying to imitate what Nature has achieved, with the precision at the nanometer level paving a way to nanobiotechnology, a division of nanotechnology that involves exploitation of bioorganic molecules on the nanoscale. The nanostructures obtained from self-assembly of bioorganic molecules are attractive due to their biocompatibility, ability for molecular recognition and ease of chemical modification, thus providing an innovative route for fabricating multifunctional bioorganic electronic devices [17]. Molecular self-assembly is the main bottom-up approach for the affordable production of bulk quantities of well-defined nanostructures. Proteins and peptides are the most versatile natural molecular building blocks, due to their extensive chemical, conformational and functional diversity [18]. Various peptide-based building blocks, such as aromatic dipeptides have been designed and developed for the construction of organized supramolecular nanostructures [19].

Diphenylalanine (FF), a fragment of the Alzheimer's β-amyloid peptide, is one of the self-assembling dipeptides that have recently been a subject of intense research in nanotechnology due to unique assembly characteristics [20]. It was found [21] that the FF self-assemble into semi-crystalline nanotubes in an aqueous solution. X-ray crystallographic analysis, first presented by Görbitz [22], showed that FF dipeptide has hexagonal crystal structure (P61 space group) and can form channel structures as well as that these simple compounds crystallize with hydrogen-bonded head-to-tail chains in the shape of helices with four or six peptide molecules per turn. The resulting structures have chiral hydrophilic channels with a van der Waals' diameter up to 10 Å. It was later shown that FF peptides can also self-assemble into diverse structures, such as nanotubes [13, 23], nanowires [24, 25], nanospheres [26], microcrystals [27, 28] or can be further designed as 0D quantum dots, depending on the deposition conditions and the type of solvent. The self-assembling mechanism, in which FF nanostructures are produced, is not yet fully understood. However, the most acceptable explanation suggests that FF self-assembling mechanism is governed by non-covalent intermolecular interactions such as electrostatic, hydrophobic, van der Waals, as well as by hydrogen bonds between aromatic rings and π-π stacking interactions [21].

The FF nanostructures are of a special interest due to their unique physical, chemical and mechanical properties. For example, FF nanotubes exhibit a high Young's modulus of $\approx$ 19 GPa [28], a huge mechanical stiffness of 160 N/m [29] and are stable under extreme conditions, including boiling water, organic solvents such as ethanol, acetone and various acidic conditions [30]. The discovery of strong piezoelectric activity [13], temperature-dependent spontaneous polarization and phase transitions in these aromatic dipeptides have established them as nanomaterials with possible ferroelectric properties [13].

3. Fabrication methods and piezoelectric properties

The self-assembly process of the FF dipeptide in tubes, as well as in rods, microribbons, fibers, among others, is carried out in the presence of liquid (mostly water) [21, 31–33]. The water molecules are contributing by electrostatic charge, allowing FF molecules to move closer to each other to start the assembly process [34]. Before mixing the original FF powder with water, it is necessary to prepare a stock solution. Many studies [35] describe a standard stock solution prepared in a concentration of 100 mg/ml in a polar solvent (acetic acid or hexafluoroisopropanol (HFIP)). Usually, to ensure the growth of a different structures, the stock solution is mixed in a different ratio with water. The standard method for preparing FF tubes is to dilute the stock solution in deionized water to a final FF concentration of 2 mg/ml [13, 21, 27]. This method produces a multitude of tubes of various lengths and diameters from tens of nm to hundreds of μm. Studies of these FF nanotubes via Piezoresponse Force Microscopy (PFM) revealed a strong piezoelectric effect, with the orientation of polarization along the tube axis [13]. The shear piezoelectric coefficient (d_{15}~60 pm/V) was obtained for the nanotubes with ~200 nm in diameter and was explained considering the dipeptide nanotubes as a crystalline nanoceramic with polar structure.

Recently, FF tubes fabricated using this method were studied in two different configurations [27]: (i) suspended ones (cantilever-type) with one edge clamped to a rigid support. For the cantilever-like configuration, the tube edge was picked up with tweezers and placed on the top of a conducting carbon tape whereas for the study in (ii) longitudinal extensional mode, the tubes were prepared directly on a patterned glass substrate coated with Au so as to comprise a conducting electrode. **Figure 1** illustrates these structures, focusing on the harvesting mode and experimental configuration.

The vibrational response was investigated using an Atomic Force Microscopy (AFM) setup where the conducting tip is used both to excite the piezoelectric vibrations and to study the vibrational response via an additional lock-in amplifier connected to the photodiode and synchronized with the driving voltage. The measurement scheme is analogous to that used in the PFM method [36] but the external ac driving was replaced with the internal one commonly used in AFMs for the cantilever resonance calibration. **Figure 2** represents the measured amplitude of the deflection (DFL) signal in respect to the excitation frequency for the cantilever

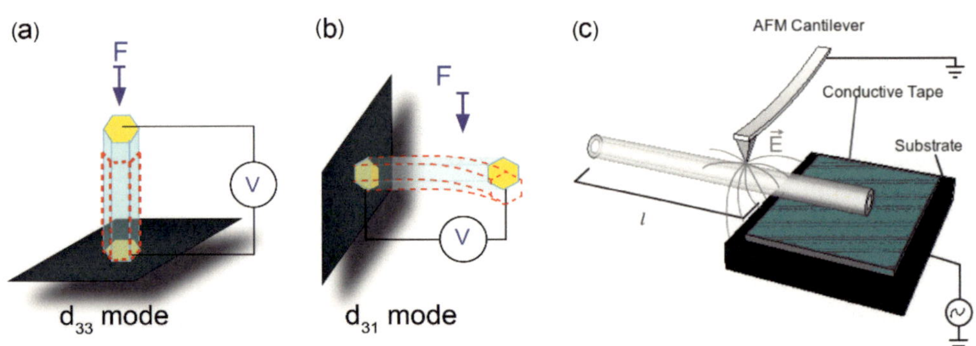

Figure 1.
Schematic of the energy harvesting configurations in (a) d_{33} and (b) d_{31} modes. (c) FF tube arrangement in the experimental situation. AFM cantilever is used to excite piezoelectric resonances and to measure induced vibrations as a function of frequency. Arrows are used to highlight inhomogeneous distribution of electric field that helps to induce bending resonances in the system. Reproduced from [27], with the permission of AIP publishing.

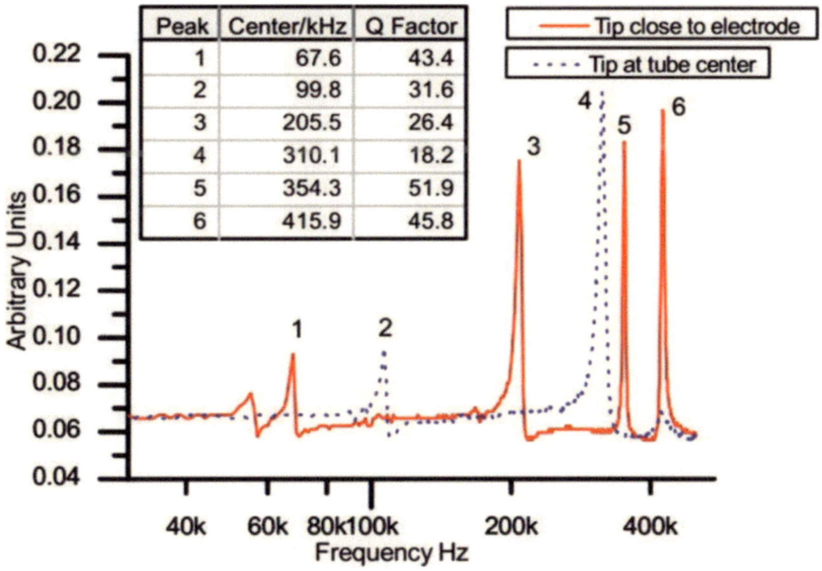

Figure 2.
Piezoelectric resonances excited in the FF microtube. Solid line represents the resonance structure for the AFM tip located in the middle. Dotted line is the example of the resonance peaks for the AFM tip close to the support. Table in the set summarizes resonance frequencies and quality factors for both configurations. Reproduced from [27], with the permission of AIP publishing.

configuration shown in **Figure 1** for two different tip positions. The studied tube was 885 μm long and 13 μm in diameter. The measured resonance frequencies and respective quality factors are shown in the inset to **Figure 2**. As it is clear from this Figure, the peaks shift to higher frequencies with increasing distance between the voltage application point (AFM tip) and the clamping point. However, the ratio between the second and the first resonance modes is approximately the same, about 3.05–3.1. The quality factors are in the range 20–50 and are close to those reported for piezoelectrically driven PVDF fibers prepared by electrospinning [37].

It was demonstrated [27] that self-assembled diphenylalanine microtubes prepared from the solution can be used as piezoelectric resonators with the potential to perform sensing and actuating functions in micro- and nanoelectromechanical systems. The sharp piezoelectric resonance in the MHz range with the quality factor > 100 was already demonstrated. These parameters can be further improved by scaling down the sizes of the devices down to sub-μm ranges.

An in-depth study of the piezoelectric properties of similar FF microtubes using the advanced PFM were reported by Vasilev *et al* [38]. Piezoelectric coefficients of the microtubes were measured in three different configurations (**Figure 3**). The full matrix of piezoelectric coefficients was also determined in this work. Piezoelectric coefficients are important input parameters for the design and fabrication of harvesting devices based on nanostructured materials, including FF peptides.

Kim *et al* [25] have shown the influence of the different ratios between the stock solution and water that impact on the hydrogen bond density, which leads to a change of the self-assembled microstructure. This fact allows to understand how the fabrication structure can be controlled. As concerned the manufacturing methods, it is important to find the deposition techniques suitable for mass production. By this way, the combination of the solution properties and deposition technique will allow forming suitable structures for the applications. Today, several liquid deposition methods can be implemented, e.g. simple drop-casting and inkjet printing technology. These will be overviewed in the following section.

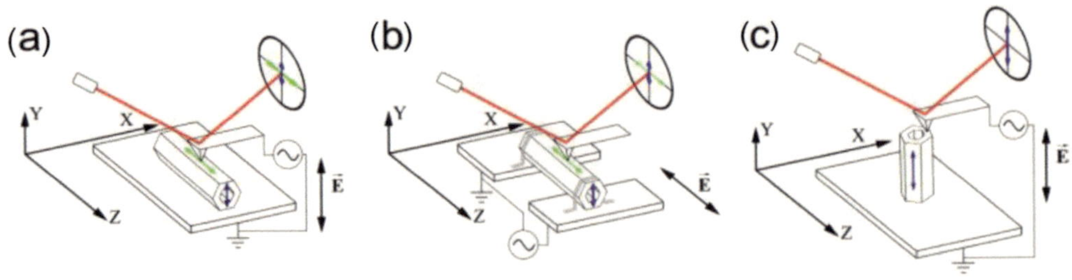

Figure 3.
Schematics of different configurations for the measurements of piezoelectric coefficients. (a) Measurement of d_{15} (LF signal, displacements are shown by horizontal arrow) and d_{14} (DFL signal, vertical arrow). (b) Measurement of clamped d_{33} (LF signal, horizontal arrow) and d_{31} (DFL signal, vertical arrow). Tube was fixed by conductive glue (c) measurement of d_{33} (DFL signal, vertical arrow). Red line shows the laser beam reflecting from the cantilever. Reprinted from [38], copyright (2016), with permission from Elsevier.

3.1 Drop casting method

The first method we discuss here is the drop casting. This method has a similar principle of operation as inkjet printing, but does not need rheology adjustment and special techniques to perform the directed growth. For the drop casting method, one just needs micropipettes, stock solution, and water. Since the crystal growth occurs in the radial direction of drying drops (from their edges to the center) on the substrate, the initial form of the droplet determines the orientation of the crystals. Thus, it is possible to control the growth direction of elongated structures. In order to implement an unidirectional growth, the method of masking with hydrophilic/hydrophilic areas was used [39]. In this study, a surface patterning technique was conducted to align FF peptide nanotubes (PNTs). Patterned regions of Si substrate were created by the growth of a silicon oxide via UV/ozone exposure through a Si mask. Further control over the diameter and density of the PNTs is achieved by heating the FF solution prior to drop casting. **Figure 4** highlights the results from solution with 2 mg/ml concentration grown on the different size of opening masks (hydrophilic regions).

This method demonstrates high-quality results by simple control of FF nanotubes growing using different wettability of patterned substrates. The modification of these aligned microtubes allows their application in microdevices.

3.2 Inkjet printing technology

Inkjet printing (IJP) of the materials is a method that allows to form the solid structure from liquid microdrops. The inkjet printing technology can be used for directing self-assembly of peptides to create controlled shapes of grown crystals, by drop geometry for useful for microelectronic applications. IJP structuring of peptides would allow depositing micro volumes with high precision on different substrates. However, the IJP method requires special sample preparation. The main parameters of materials for IJP are rheology, surface tension, pH and particle size. As was described above, the standard method for self-assembly FF molecules and microstructure growth is a mixing of stock solution and water. Since this mixture forms a precipitate, it cannot be used directly for the printing method. Recently, an ink composition suitable for printing and allowing one to self-assemble the structure has been developed [40]. This work demonstrated the controlled deposition of FF using drop-on-demand piezoelectric printing technology combined with high resolution and reproducibility (**Figure 5a**). A new conformation of the printed FF-based

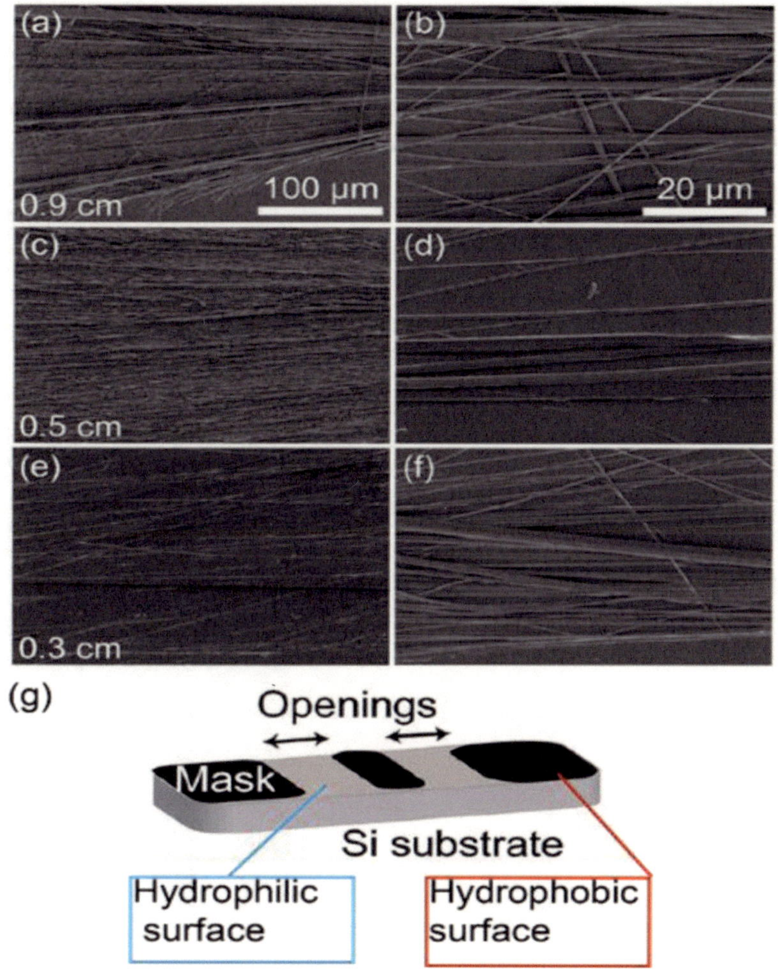

Figure 4.
SEM images grown tubes on the different size of opening masks: (a, b)—0.9 cm; (c, d)—0.5 cm; (e, f)—0.3 cm; (g)—Schematic of Si substrate preparation: Opening regions treated by UV/ozone for good wettability. Reproduced from ref. 39 with permission from the Royal Society of Chemistry.

crystals grown from the ink, which was developed for this deposition method was also demonstrated (**Figure 5b**). As is well known [40], printing ink must have certain parameters for successful application. Viscous solvents and surfactants are commonly used to change the viscosity and surface tension. But how do they affect the self-assembled structure and its properties? Nowadays, it is known that solvents act on FF molecules and cause significant electrostatic interactions between them [21]. This interaction allows the formation of various microstructures, which is quite evident as exemplified by the methanol – water mixture for the growth FF crystals [41].

In the case of ethylene glycol and surfactant, a change of grown microstructure was noticed (**Figure 5c**). That should mean the change in the self-assembly process. As a result, a new crystalline structure was formed: this is not a hexagonal tube, but a flat rhombic belt. The piezoresponse of that structure was investigated by PFM method. The obtained FF belt-like crystals exhibited a high piezoelectric response (**Figure 5d**), a piezoelectric coefficient d_{eff} of 40 ± 5 pm/V; this value is smaller than the one observed for typical hexagonal FF microtubes (60 ± 10 pm/V) [13], but it is still higher than those measured in many organic piezoelectrics including polyvinylidene fluoride [42]. The value of the piezoelectric coefficient allows the use this material for microdevice applications. In addition, the solution was adapted to IJP technology which opened new possibilities for the application and fabrication functional structure on the different substrates. In contrast to the simple drop cast method, IJP allows depositing microstructures with controlled self-assembly order.

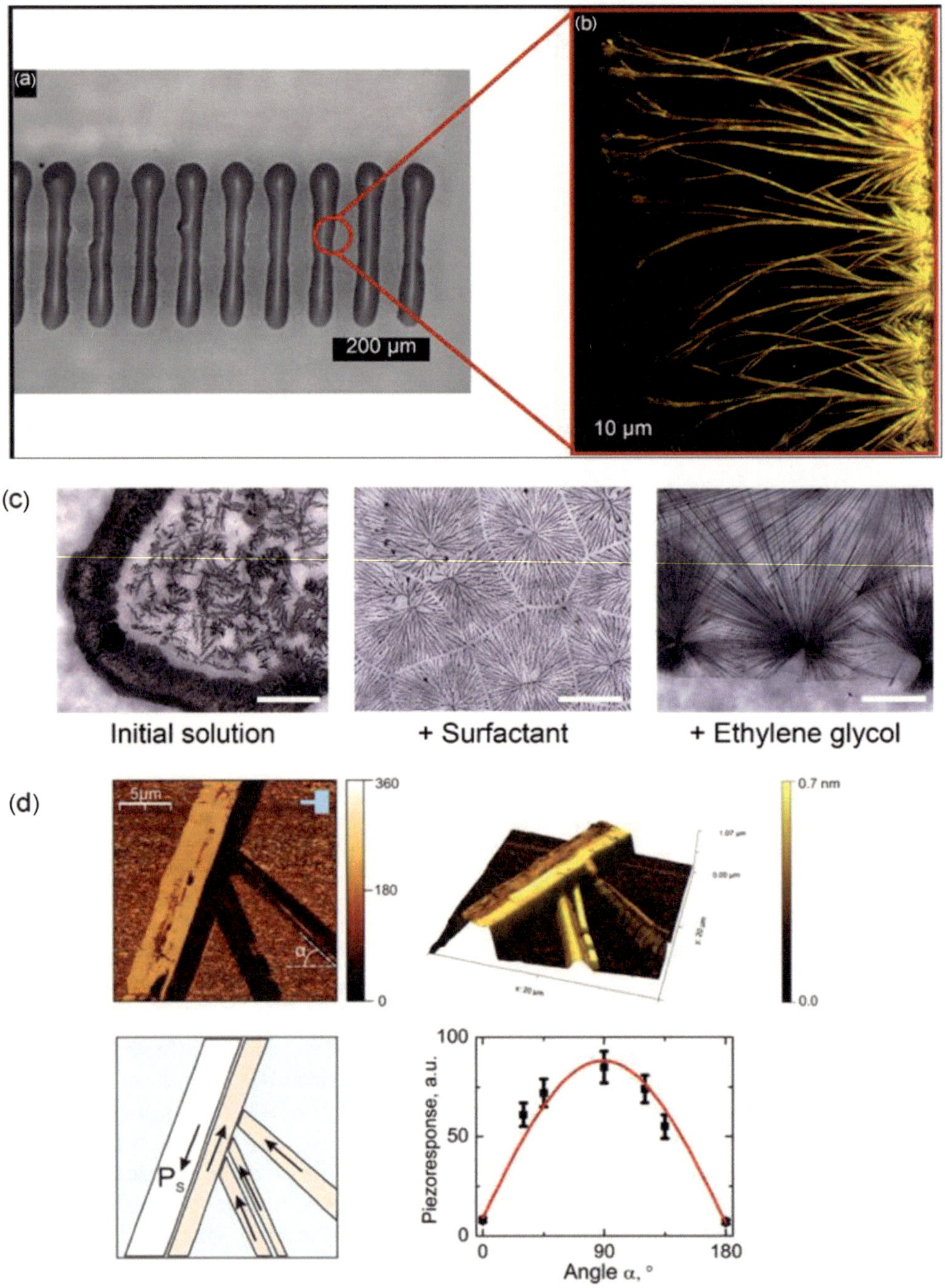

Figure 5.
Optical and PFM investigation of printed FF structure: (a, b) inkjet printed FF crystals grown orthogonal to the drop edge; (c) influence of the additives on the microstructure FF crystals (scale bar 500 μm); (d) PFM measurement of printed crystals. Reprinted with permission from [40]. Copyright © 2018 American Chemical Society.

3.3 Dip coating technology

The dip-coating method is a common technique for the material deposition from liquids. This method allows fabricating thin and thick films by dipping and pulling the substrate in the bath with the solution. Since this method allows to control the velocity of pulling and dipping, —it can be used for crystal growth and it was suggested to align FF nanotubes on the polymer substrate (**Figure 6a**) [43].

By controlling the stock solution concentration in water and pulling speed, the morphology of FF peptide nanostructures from the vesicles to elongated and aligned nanotubes was found (**Figure 6b**). Furthermore, FF nanotubes show unipolarity (**Figure 6c**), which is critical for the fabrication as piezoelectric energy harvesters as nanogenerators.

3.4 Benefits and limitations of methods

Before comparing the discussed methods, the application areas must be first defined. If the device to be designed does not require directed growth and the cost should be low, the drop-casting method will be enough, because it is the simplest technique and cost-efficient. However, if the device should be anisotropic and directional response is needed drop casting on hydrophilic and hydrophobic surfaces is required [39]. By the puling and dipping of substrate (dip-coating method), the deposition of oriented tubes on many surfaces can be performed, although it depends on wettability of the substrates. But two big limitations of dip-coating are: (1) both sides of the substrate are covered by tubes (2) deposition occurs

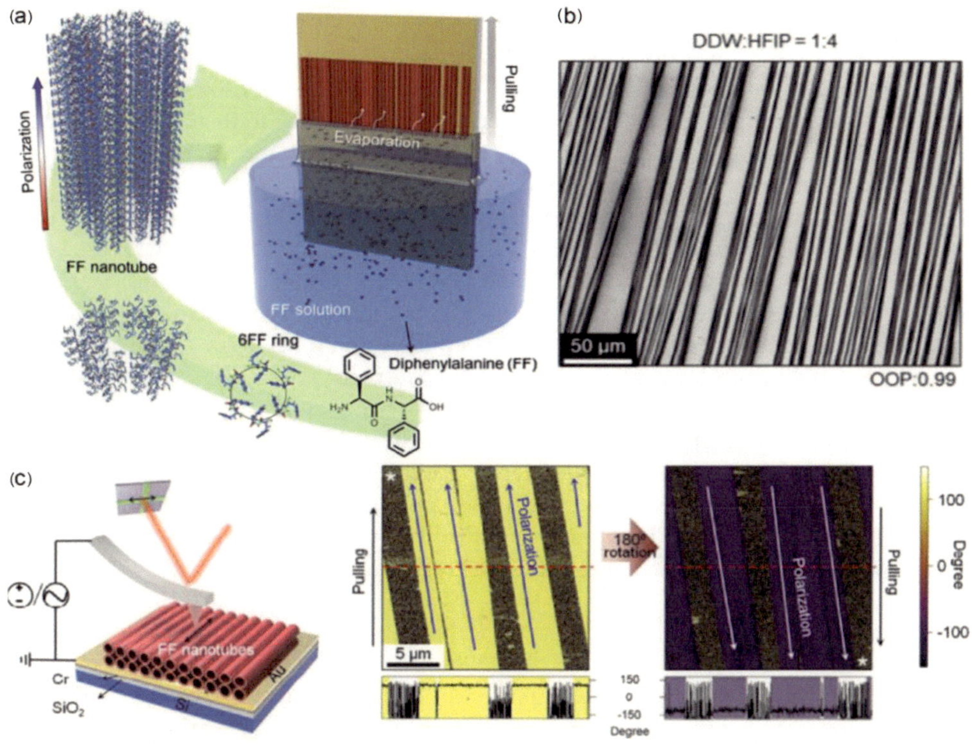

Figure 6.
Dip-coated aligned FF nanotubes: (a)—Schematic of the substrate pulling from a mixture of FF stock solution and water: (b)—SEM image of the aligned FF tubes: (c)—PFM measurement demonstrated one direction of polarization in parallel tubes. Reprinted with permission from [43]. Copyright © 2018 American Chemical Society.

everywhere and additional patterning will be required. For the directed deposition, the most effective method is inkjet printing, which allows jet drop the volumes as small as 10 pL. The complexity of this method is to find an appropriate solution formulation. A stock solution, mixed with many solvents, is demanded to prepare a solution with required rheology, so that the microstructure can be adjusted to specific application [40].

4. Examples of energy harvesting applications

4.1 Nanogenerator

The discovery of ZnO-based nanogenerators in 2006 [44] significantly boosted the research on piezoelectric materials for energy harvesting applications. Piezoelectric nanogenerators are based on the piezoelectric effect through which mechanical energy is converted into electricity and vice versa via linear coupling of mechanical strain and electric field. Since the discovery of piezoelectric nanogenerators based on ZnO, many other nanogenerators have also been developed, e.g. based on flexible PVDF [45–47], triboelectric [48–50] and pyroelectric ones [6–8].

Peptide-based biomaterials were also recognized suitable for electrical energy conversion. Recently, it has been reported [51] that FF vertical microrod arrays grown under an external applied electric field, exhibit an effective piezoelectric constant d33 of 17.9 pm/V The effect of the electric field resulted in a stretching of the peptide backbone with a concurrent increase of the dipole moment. A power generator was developed based on vertical FF microrod arrays with uniform polarization by applying electric fields (**Figure 7a, b**) [51]. The FF-based power generator provided an open-circuit voltage of 1.4 V with an applied force of 60 N (**Figure 7c, d**), and a power density of 3.3 nW cm^{-2} at 50 MΩ, which is about 4 times higher than the energy harvested by zinc oxide nanowires [44]. The FF-based nanogenerator was integrated with a single-electrode triboelectric nanogenerator to form a hybrid device [52]. Hybrid nanogenerator combines the advantages of both piezoelectric and triboelectric nanogenerators, being able to provide the output voltages of 2.2 V.

Recently, a meniscus-driven self-assembly process was used to fabricate horizontal large-scale FF microtube arrays with an asymmetric shape (**Figure 8a**) [43]. The FF nanostructure was adjusted by the type of the solvent, the solubility and pulling speed. The resulting FF nanotube devices produce electrical energy upon application of mechanical force. The FF-based device exhibited a voltage of 2.8 V, a current of 37.4 nA and a power of 8.2 nW, respectively, when a force of 42 N was applied [43] and can power multiple liquid-crystal display (LCD) panels and can power multiple liquid-crystal display panels (**Figure 8b, c**). These peptide-based energy-harvesting materials are considered as compatible energy source for biomedical applications in a recent future. The parameters of several types of piezoelectric nanogenerators are compared in **Table 1**. It can be seen that piezoelectric nanogenerator based on horizontal FF tubes is the most efficient one.

4.1.1 Energy storage

Piezoelectric biomaterials as exemplified by nanostructured peptides can be also used for supercapacitors, due to their large specific surface area. Vapor deposition technology has been recently used to grow FF peptide nanotubes arrays as a part of electric double layer capacitor (EDLC) [55]. The capacitor based on these nanotubes asmodified electrode showed a high areal capacitance of 480 μF cm^{-2} which was is

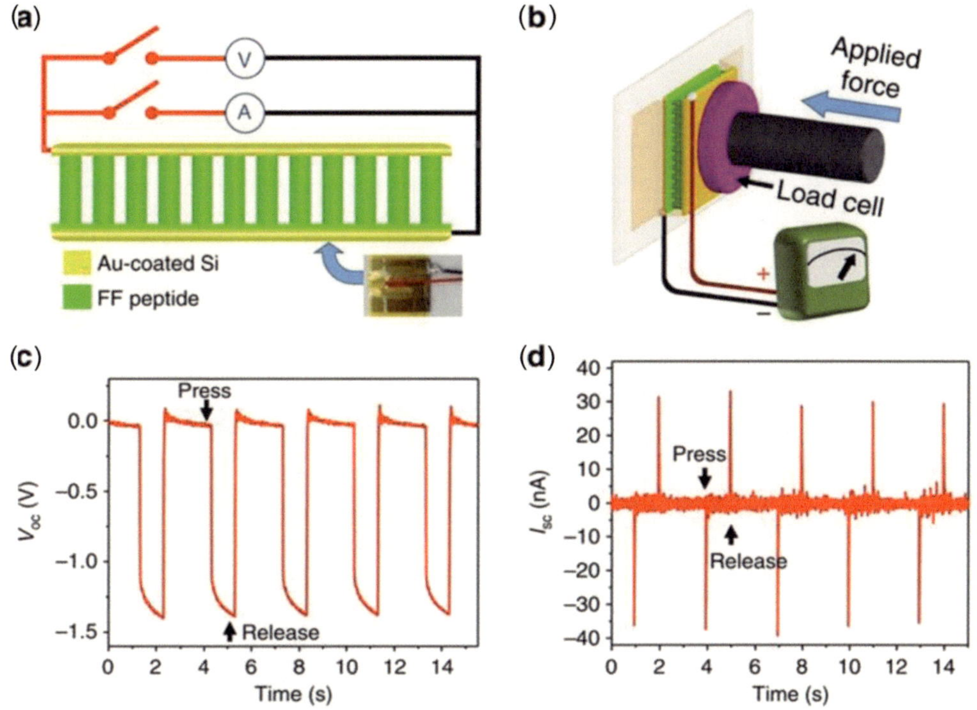

Figure 7.
(a) Schematic of the FF peptide-based generator connected to the measurement equipment. Bottom-right inset: Photography of a real device. (b) Schematic of the measurement set-up in which a linear motor pushes with controlled forces on the top electrode in (a). The linear motor was programmed to always keep contact with the top electrode to minimize the effect of static charges. (c, d) open-circuit voltage (c) and short-circuit current (d) from a generator using microrods from positive-EF growth. Reproduced from [51] with the permission.

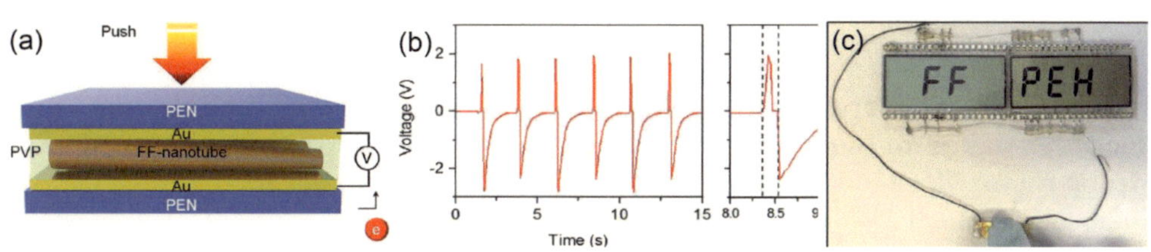

Figure 8.
(a) Schematic illustration of FF peptide-based piezoelectric energy harvester (PEH); (b) Output voltage responses with 1 GΩ external load; (c) Photograph of the LCD driven by pressing the PEH with a finger. Reprinted with permission from [43]. Copyright © 2018 American Chemical Society.

four times as high as that for carbon nanotubes modified electrodesand 30 times higher than that for carbon electrodes without any modification. The FF nanotubes-based capacitor demonstrated very high cycle stability and no significant capacity loss after 10,000 charge–discharge cycles [55]. An external electric field was applied to generate unidirectionally aligned and stable FF nanotube/microtube arrays at room temperature [56]. The FF nanotube/microtubes with open ends morphology and multi-layer walls were considered as the origin of the large specific surface area. The FF nanostructures were used as electrodes to fabricate a supercapacitor, and a high specific capacity of 1000 μF cm^{-2} at a scan rate of 50 mV s^{-1} was obtained (**Figure 9**).

Generator	V_{OC}, V	Energy density	d_{33}, pm/V	Applied force, N
PENG based on ZnO [53]	1.26	2.7 nW/cm^3	12.0 [54]	—
PENG based on vertical FF [51]	1.4	3.3 nW/cm^2	17.9	60
PENG based on horizontal FF [43]	2.8	8.2 nW	—	42

Table 1.
Parameters of piezolectric (PENG) nanogenerators.

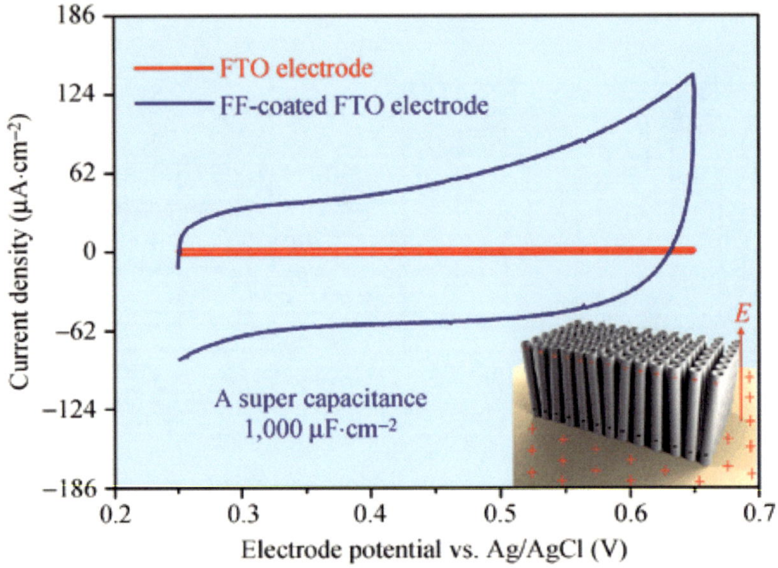

Figure 9.
Cyclic voltammetry measurements of the FTO electrode coated with unidirectionally aligned NTs/MTs (red line). Reprinted by permission from [56]. Springer nature. Copyright © 2014.

It has been reported [14, 55, 57] that FF nanostructures hold a potential as high-capacity ultracapacitors for fast charging in energy storage applications such as batteries for mobile devices and electric vehicles. In particular, vertical FF nanostructures can increase the effective activation area of modified electrodes and increase electrical conductivity [55].

5. Conclusions

The purpose of this book chapter was to give an overview of the most relevant findings in the piezoelectric studies on self-assembled peptides, in particular using diphenylalanine. Diphenylalanine, as a piezoelectric self-assembled dipeptide, is important from both the fundamental point of view and real-world energy harvesting applications. The morphology of FF peptide, the mechanism of self-assembly, the doping/composite structure are of particular interest due to their effect on piezoelectricity From the point of view of applications, two major applications were highlighted: nanogenerators based on piezoelectric biomaterials, such as FF peptides, and energy storage capacitors. Both types of nanogenerators described in this Chapter combine high efficiency, stability and self-assembled features for energy harvesting and storage that are desirable for future applications. Further, the research on self-assembled materials with a high piezoelectric response, by controlling their composition and structure, can lead to the next generation of nanogenerators in a near future.

Acknowledgements

This work was developed within the scope of the project CICECO-Aveiro Institute of Materials, FCT ref. UID/CTM/50011/2019, financed by national funds through the FCT/MCTES. S.K. and A.K. were partly supported by FCT (Portugal) through the Project PTDC/CTM-CTM/31679/ 2017 - "BioPiezo" and joint Portugal-Turkey project (TUBITAK/0006/2014). M. Soares dos Santos was also supported by FCT, through the grant references SFRH/BPD/117475/2016. All authors were partly supported by FCT through the project "SelfMED" (POCI-01-0145-FEDER-031132). Part of this work was funded by national funds (OE), through FCT – Fundação para a Ciência e a Tecnologia, I.P., in the scope of the framework contract foreseen in the numbers 4, 5, and 6 of the article 23, of the Decree-Law 57/2016, of August 29, changed by Law 57/2017, of July 19.

Conflict of interest

The authors declare no conflict of interest.

Author details

Vladislav Slabov[1], Svitlana Kopyl[1], Marco P. Soares dos Santos[2,3] and Andrei Kholkin[1,4]*

1 Department of Physics and CICECO-Aveiro Institute of Materials, University of Aveiro, Aveiro, Portugal

2 Department of Mechanical Engineering and Centre for Mechanical Technology and Automation (TEMA), University of Aveiro, Aveiro, Portugal

3 Associated Laboratory for Energy, Transports and Aeronautics (LAETA), Porto, Portugal

4 School of Natural Sciences and Mathematics, Ural Federal University, Ekaterinburg, Russian Federation

*Address all correspondence to: kholkin@ua.pt

References

[1] Priya S, Inman DJ, editors. Energy Harvesting Technologies. US: Springer; 2009. p. 524. DOI: 10.1007/978-0-387-76464-1

[2] Eliseev EA, Morozovska AN, Glinchuk MD, Zaulychny BY, Skorokhod VV, Blinc R. Surface-induced piezomagnetic, piezoelectric, and linear magnetoelectric effects in nanosystems. Physical Review B. 2010;**82**:085408. DOI: 10.1103/PhysRevB.82.085408

[3] Gouveia PJ, Rosa S, Ricotti L, Abecasis B, Almeida HV, Monteiro L, et al. Flexible nanofilms coated with aligned piezoelectric microfibers preserve the contractility of cardiomyocytes. Biomaterials. 2017;**139**:213-228. DOI: 10.1016/j.biomaterials.2017.05.048

[4] Soares dos Santos MP, Coutinho J, Marote A, Sousa B, Ramos A, Ferreira JAF, et al. Capacitive technologies for highly controlled and personalized electrical stimulation by implantable biomedical systems. Scientific Reports. 2019;**9**:5001. DOI: 10.1038/s41598-019-41540-3

[5] Li X, Lu S-G, Chen X-Z, Gu X, Qian X-S, Zhang QM. Pyroelectric and electrocaloric materials. Journal of Materials Chemistry C. 2001;**1**:23-37. DOI: 10.1039/C2TC00283C

[6] Wang ZL, Zhu G, Yang Y, Wang S, Pan C. Progress in nanogenerators for portable electronics. Materials Today. 2012;**15**:532-543. DOI: 10.1016/S1369-7021(13)70011-7

[7] Lee J-H, Ryu H, Kim T-Y, Kwak S-S, Yoon H-J, Kim T-H, et al. Thermally induced straincoupled highly stretchable and sensitive pyroelectric nanogenerators. Advanced Energy Materials. 2015;**5**:1500704. DOI: 10.1002/aenm.201500704

[8] Yang Y, Wang S, Zhang Y, Wang Z. Pyroelectric nanogenerators for driving wireless sensors. Nano Letters. 2012;**12**:6408-6413. DOI: 10.1021/nl303755m

[9] Lines ME, Glass AM. Principles and Applications of Ferroelectrics and Related Materials. 1 st ed. Oxford, England: Clarendon Press; 1977. pp. 559-604. DOI: 10.1093/acprof:oso/9780198507789.001.0001

[10] Cardoso VF, Correia DM, Ribeiro C, Fernandes MM, Lanceros-Méndez S. Fluorinated polymers as smart materials for advanced biomedical applications. Polymers. 2018;**10**:161. DOI: 10.3390/polym10020161

[11] Hutmacher DW. Scaffolds in tissue engineering bone and cartilage. Biomaterials. 2000;**21**:2529-2543. DOI: 10.1016/S0142-9612(00)00121-6

[12] Guerin S, Stapleton A, Chovan D, Mouras R, Gleeson M, McKeown C, et al. Control of piezoelectricity in amino acids by supramolecular packing. Nature Materials. 2018;**17**:180-186. DOI: 10.1038/nmat5045

[13] Kholkin AL, Amdursky N, Bdikin I, Gazit E, Rosenman G. Strong piezoelectric activity in peptide nanotubes. ACS Nano. 2010;**4**:610-614. DOI: 10.1021/nn901327v

[14] Tao K, Makam P, Aizen R, Gazit E. Self-assembling peptide semiconductors. Science. 2017;**358**:6365. DOI: 10.1126/science.aam9756

[15] Minary-Jolandan M, Yu MF. Nanoscale characterization of isolated individual type I collagen fibrils: Polarization and piezoelectricity. Nanotechnology. 2009;**20**:085706. DOI: 10.1088/0957-4484/20/8/085706

[16] Yuan H, Lei T, Qin Y, He J-H, Yang R. Design and application of piezoelectric biomaterials.

Journal of Physics D: Applied Physics. 2019;**52**:194002. DOI: 10.1088/1361-6463/ab0532

[17] Lee JS, Yoon I, Kim J, lhee H, Kim B, Park CB. Self-assembly of semiconducting photoluminescent peptide nanowires in the vapor phase. Angewandte Chemie. 2011;**123**:1196-1199. DOI: 10.1002/anie.201003446

[18] Bystrov VS, Bdikin I, Heredia A, Pullar RC, Mishina E, Sigov AS, Kholkin AL. Piezoelectricity and ferroelectricity in biomaterials: From proteins to self-assembled peptide nanotubes. In: Piezoelectric Nanomaterials for Biomedical Applications. Ciofani G, Menciassi A, editors. Berlin Heidelberg: Springer-Verlag; 2012. pp. 187-211. DOI: 10.1007/978-3-642-28044-3_7

[19] Huang RL, Wu SK, Li AT, Li Z. Integrating interfacial self-assembly and electrostatic complexation at an aqueous interface for capsule synthesis and enzyme immobilization. Journal of Materials Chemistry A. 2014;**2**:1672-1676. DOI: 10.1186/1556-276X-9-653

[20] Cipriano T, Knotts G, Laudari A, Bianchi RC, Alves WA, Guha S. Bioinspired peptide nanostructures for organic field-effect transistors. ACS Applied Materials & Interfaces. 2014;**6**:21408. DOI: 10.1021/am5064124

[21] Reches M, Gazit E. Casting metal nanowires within discrete self-assembled peptide nanotubes. Science. 2003;**300**:625. DOI: 10.1126/science.1082387

[22] Görbitz CH. Nanotube formation by hydrophobic dipeptides. Chemistry - A European Journal. 2001;7:5153-5159. DOI: 10.1002/1521-3765(20011203)7:23<5153:AID-CHEM5153>3.0.CO;2-N

[23] Bdikin I, Bystrov V, Kopyl S, Lopes R, Delgadillo I, Gracio J, et al. Evidence of ferroelectricity and phase transition in pressed diphenylalanine peptide nanotubes. Applied Physics Letters. 2012;**100**:043702. DOI: 10.1063/1.3676417

[24] Sasso L, Vedarethinam I, Emnéus J, Svendsen WE, Castillo-León J. Self-assembled diphenylalanine nanowires for cellular studies and sensor applications. Journal of Nanoscience and Nanotechnology. 2012;**12**(4):3077-3083. DOI: 10.1166/jnn.2012.4534

[25] Kim J, Han TH, Kim Y-I, Park JS, Choi J, Churchill DG, et al. Role of water in directing diphenylalanine assembly into nanotubes and nanowires. Advanced Materials. 2010;**22**:583-587. DOI: 10.1002/adma.200901973

[26] Guo C, Luo Y, Zhou R, Wei G. Probing the self-assembly mechanism of Diphenylalanine-based peptide Nanovesicles and nanotubes. ACS Nano. 2012;**65**:3907-3918. DOI: /10.1021/nn300015g

[27] Bosne ED, Heredia A, Kopyl S, Karpinsky DV, Pinto AG, Kholkin AL. Piezoelectric resonators based on self-assembled diphenylalanine microtubes. Applied Physics Letters. 2013;**102**:073504. DOI: 10.1063/1.4793417

[28] Nikitin T, Kopyl S, Shur VY, Kopelevich YV, Kholkin AL. Low-temperature photoluminescence in self-assembled diphenylalanine microtubes. Physics Letters A. 2016;**380**:1658-1662. DOI: 10.1016/j.physleta.2016.02.043

[29] Kol N, Adler-Abramovich L, Barlam D, Shneck RZ, Gazit E, Rousso I. Self-assembled peptide nanotubes are uniquely rigid bioinspired supramolecular structures. Nano Letters. 2005;**5**:1343-1346. DOI: 10.1021/nl0505896

[30] Adler-Abramovich L, Reches M, Sedman VL, Allen S, Tendler SJB, Gazit E. Thermal and chemical stability

of Diphenylalanine peptide nanotubes: Implications for nanotechnological applications. Langmuir. 2006;**22**:1313. DOI: 10.1021/la052409d

[31] Nuraeva A, Vasilev S, Vasileva D, Zelenovskiy P, Chezganov D, Esin A, et al. Evaporation-driven crystallization of Diphenylalanine microtubes for microelectronic applications. Crystal Growth & Design. 2016;**16**:1472-1479. DOI: 10.1021/acs.cgd.5b01604

[32] Levin A, Mason TO, Adler-Abramovich L, Buell AK, Meisl G, Galvagnion C, et al. Ostwald's rule of stages governs structural transitions and morphology of dipeptide supramolecular polymers. Nature Communications. 2014;**13**:5219. DOI: 10.1038/ncomms6219

[33] Wang M, Du L, Wu X, Xiong S, Chu PK. Charged Diphenylalanine nanotubes and controlled hierarchical self-assembly. ACS Nano. 2011;**5**: 4448-4454. DOI: 10.1021/nn2016524

[34] Ciferri A, Perico A. Ionic Interactions in Natural and Synthetic Macromolecules. Hoboken, New Jersey: John Wiley & Sons Inc.; 2012. p. 872. DOI: 10.1002/9781118165850

[35] Mason TO, Chirgadze DY, Levin A, Adler-Abramovich L, Gazit E, Knowles TPJ, et al. Expanding the solvent chemical space for self-assembly of dipeptide nanostructures. ACS Nano. 2014;**8**:1243-1253. DOI: 10.1021/nn404237f

[36] Balke N, Bdikin IK, Kalinin SV, Kholkin AL. Electromechanical imaging and spectroscopy of ferroelectric and piezoelectric materials: State of the art and prospects for the future. Journal of the American Ceramic Society. 2009;**92**:1629-1647. DOI: 10.1111/j.1551-2916.2009.03240.x

[37] Urwyler P, Schif H, Gobbrect J, Hafeli O, Altana M, Battiston F, et al. Surface patterned polymer micro-cantilever arrays for sensing. Sensors and Actuators, A: Physical. 2011; **172**:2-8. DOI: 10.1016/j.sna.2010.12.007

[38] Vasilev S, Zelenovskiy P, Vasileva D, Nuraeva A, Shur VY, Kholkin AL. Piezoelectric properties of diphenylalanine microtubes prepared from the solution. Journal of Physics and Chemistry of Solids. 2016;**93**:68-72. DOI: /10.1016/j.jpcs.2016.02.002

[39] Almohammed S, Oladapo SO, Ryan K, Kholkin AL, Rice JH, Rodriguez BJ. Wettability gradient-induced alignment of peptide nanotubes as templates for biosensing applications. RSC Advances. 2016;**6**:41809-41815. DOI: 10.1039/C6RA05732B

[40] Safaryan S, Slabov V, Kopyl S, Romanyuk K, Bdikin I, Vasilev S, et al. Diphenylalanine-based microribbons for piezoelectric applications via inkjet printing. ACS Applied Materials & Interfaces. 2018;**10**:10543-10551. DOI: 10.1021/acsami.7b19668

[41] Zelenovskiy PS, Shur VY, Nuraeva AS, Vasilev SG, Vasileva DS, Alikin DO, et al. Morphology and piezoelectric properties of Diphenylalanine microcrystals grown from methanol-water solution. Ferroelectrics. 2015;**475**:127-134. DOI: 10.1080/00150193.2015.995577

[42] Ramadan KS, Sameoto D, Evoy S. A review of piezoelectric polymers as functional materials for electromechanical transducers. Smart Materials and Structures. 2014;**23**:033001. DOI: 10.1088/0964-1726/23/3/033001

[43] Lee J-H, Heo K, Schulz-Schönhagen K, Lee JH, Desai MS, Jin H-E, et al. Diphenylalanine peptide nanotube energy harvesters. ACS Nano. 2018;**12**:8138-8144. DOI: 10.1021/acsnano.8b03118

[44] Wang ZL, Song J. Piezoelectric nanogenerators based on zinc oxide nanowire arrays. Science. 2006;**312**: 242-246. DOI: 10.1126/science.1124005

[45] Yan J, Liu M, Jeong lG, Kang W, Li L, Zhao Y, et al. Performance enhancements in poly(vinylidene fluoride)-based piezoelectric nanogenerators for efficient energy harvesting. Nano Energy. 2019;**56**:662-692. DOI: 10.1016/j.nanoen.2018.12.010

[46] Hu D, Yao M, Fan Y, Ma C, Fan M, Liu M. Strategies to achieve high performance piezoelectric nanogenerators. Nano Energy. 2019;**55**:288-304. DOI: 10.1016/j.nanoen.2018.10.053

[47] Fan FR, Tang W, Wang ZL. Flexible Nanogenerators for energy harvesting and self-powered electronics. Advanced Materials. 2016;**28**:4283-4305. DOI: 10.1002/adma.201504299

[48] Chen J, Zhu G, Yang W, Jing Q, Bai P, Yang Y, et al. Harmonic-resonator-based triboelectric nanogenerator as a sustainable power source and a self- powered active vibration sensor. Advanced Materials. 2013;**25**:6094-6099. DOI: 10.1002/adma.201302397

[49] Niu S, Wang S, Lin L, Liu Y, Zhou YS, Hu Y, et al. Theoretical study of contact-mode triboelectric nanogenerators as an effective power source. Energy & Environmental Science. 2013;**6**:3576-3583. DOI: 10.1039/C3EE42571A

[50] Cheng L, Xu Q, Zheng Y, Jia X, Qin Y. A selfimproving triboelectric nanogenerator with improved charge density and increased charge accumulation speed. Nature Communications. 2018;**9**:3773. DOI: 10.1038/s41467-018-06045-z

[51] Nguyen V, Zhu R, Jenkins K, Yang R. Self-assembly of diphenylalanine peptide with controlled polarization for power generation. Nature Communications. 2016;**7**:13566. DOI: 10.1038/ncomms13566

[52] Vu N, Kelly S, Yang R. Piezoelectric peptide-based nanogenerator enhanced by single-electrode triboelectric nanogenerator. APL Materials. 2017;**5**:074108. DOI: 10.1063/1.4983701

[53] Xu S, Qin Y, Xu C, Wei Y, Yang R, Wang ZL. Self-powered nanowire devices. Nature Nanotechnology. 2010;**5**:366-373. DOI: 10.1038/nnano.2010.46

[54] Bdikin IK, Gracio J, Ayouchi R, Schwarz R, Kholkin AL. Local piezoelectric properties of ZnO thin films prepared by RF-plasma-assisted pulsed-laser deposition method. Nanotechnology. 2010;**21**:235703. DOI: 10.1088/0957-4484/21/23/235703

[55] Adler-Abramovich L, Aronov D, Beker P, Yevnin M, Stempler S, Buzhansky L, et al. Self-assembled arrays of peptide nanotubes by vapour deposition. Nature Nanotechnology. 2009;**4**:849-854. DOI: 10.1038/nnano.2009.298

[56] Zhang J, Wu X, Gan Z, Zhu X, Jin Y. Unidirectionally aligned diphenylalanine nanotube/ microtube arrays with excellent supercapacitive performance. Nano Research. 2014;**7**:929-937. DOI: 10.1007/s12274-014-0455-6

[57] Beker P, Koren I, Amdursky N, Gazit E, Rosenman G. Bioinspired peptide nanotubes as supercapacitor electrodes. Journal of Materials Science. 2010;**45**:6374-6378. DOI: 10.1007/s10853-010-4624-z

Section 3

Nanogenerator for Biomedical Applications

Chapter 6

Piezoelectric/Triboelectric Nanogenerators for Biomedical Applications

Panpan Li, Jeongjae Ryu and Seungbum Hong

Abstract

Bodily movements can be used to harvest electrical energy via nanogenerators and thereby enable self-powered healthcare devices. In this chapter, first we summarize the requirements of nanogenerators for the applications in biomedical fields. Then, the current applications of nanogenerators in the biomedical field are introduced, including self-powered sensors for monitoring body activities; pacemakers; cochlear implants; stimulators for cells, tissues, and the brain; and degradable electronics. Remaining challenges to be solved in this field and future development directions are then discussed, such as increasing output performance, further miniaturization, encapsulation, and improving stability. Finally, future outlooks for nanogenerators in healthcare electronics are reviewed.

Keywords: nanogenerator, biomedical application, self-powered healthcare devices, energy harvester

1. Introduction

The ongoing development of nanogenerators in recent years has enabled the design of self-powered systems that can operate without external power supplies. Nanogenerators have the ability to harvest mechanical energy in different forms from a variety of sources, including human body motion and activities. This makes them particularly suitable for applications in the biomedical field. Nanogenerators can convert the tiny mechanical energy in body motion, muscle contraction/relaxation, bone strain, and respiration into electrical energy [1–4]. The generated electrical energy can be used as a sustainable energy source for implantable biomedical devices, which would both reduce the volume of the powering unit and eliminate the need for battery replacement [5–7].

A great deal of work has been invested in the study of biomedical applications of nanogenerators, including self-powered sensors, pacemakers, and stimulators, and the results have shown that nanogenerators can be very promising in the biomedical field [8–15].

In this chapter, we first introduce the required characteristics of nanogenerator materials that can be used in the biomedical field. Generally, there are two main types of biomedical nanogenerators, piezoelectric nanogenerators (PENG) and triboelectric nanogenerators (TENG), which have different operating mechanisms. PENG are based on piezoelectric materials, such as polyvinylidene fluoride (PVDF) [8], poly(vinylidenefluoride-co-trifluoroethylene) [P(VDF-TrFE)] [9], $BaTiO_3$ (BTO) [10], ZnO [11], $Pb(Zr_xTi_{1-x})O_3$ (PZT) [12], and $(1 - x)Pb(Mg_{1/3}Nb_{2/3})O_3$-$xPbTiO_3$ (PMN-PT) [13]. While TENG are based on triboelectric charges which are generated when dissimilar materials are in contact [14, 15], their operating mechanism is a combination of tribo-electrification and electrostatic induction between the two contacted materials [14, 15]. A broad range of materials exhibiting these effects can be selected, which make TENG ideal for biomedical applications. Besides the PENG and TENG nanogenerators, there are also other types of biomedical nanogenerators using biofuel cells (BFCs) or photovoltaics. BFCs transform chemical energy into electrical energy from molecules present in human body [16], which are very promising since there is >100 W of chemical energy in our body [17]. Flexible photovoltaic materials can meet the conformability requirements of e-skin, thus showing the possibility of solar-powered e-skin [18, 19].

Next, we will provide some examples of important biomedical nanogenerator applications, including self-powered human activity sensors; pacemakers; cochlear implants; simulators for cells, tissues, and brain; and biodegradable electronics. After that, we will also discuss challenges and future outlooks for biomedical nanogenerators, including their miniaturization, stability, encapsulation, and output performance. We hope this book chapter will provide insight and inspiration to people who are interested in biomedical devices and nanogenerator development.

2. Nanogenerator materials for biomedical applications

Self-powered biomedical devices require nanogenerators that can directly harvest energy from their surroundings, in this case, from activities in the human body. This also requires the nanogenerators to have specific designs that respond to different mechanical stimuli with high sensitivity, since many bodily activities are subtle.

The materials used in biomedical nanogenerators should also be biocompatible. The primary conventional piezoelectric material is lead zirconate titanate (PZT). PZT has a high piezoelectric coefficient; however, the toxicity of Pb makes it unsuitable for application in the human body. Scientists have been searching for other materials in efforts to develop alternatives to lead-based nanogenerators. One of the emerging lead-free piezoelectric materials, $0.5Ba(Zr_{0.2}Ti_{0.8})O_3$-$0.5(Ba_{0.7}Ca_{0.3})TiO_3$ (BZT-BCT), has a piezoelectric coefficient comparable to PZT and also good biocompatibility, which makes it a promising candidate for applications in the biomedical field [10]. ZnO has also attracted great interest because of its favorable characteristics, which include piezoelectricity, biocompatibility, transparency, and large-area fabrication [11].

In many cases, nanogenerators based on nanowires, nanobelts, and nanorods can be placed into specific structures to fit inside the body. Nanostructures, nanocomposites, or piezoelectric polymers specifically designed with superior flexibility and elasticity are particularly preferred for biomedical applications. For example, poly(vinylidenefluoride-co-trifluoroethylene) [P(VDF-TrFE)]-based nanogenerators have demonstrated good piezoelectric coefficient, flexibility, and biocompatibility [20–24].

Finally, nanogenerators used in the biomedical field should have high sensitivity and efficiency because many bodily activities, such as respiration, heartbeat, muscle stretching, or blood circulation, are very gentle and render a small amplitude. Nanogenerators need high energy conversion efficiency and sufficient output power to be used in devices with comparable size [25–27].

Piezoelectric materials	Characteristics	Biomedical applications
PZT	High piezoelectric coefficient, toxicity	Energy harvesting from body motion, including the heart, lung, and diaphragm [4, 28] Cochlear implant [29] Eye fatigue detection [30]
PMN-PT	High piezoelectric coefficient, toxicity	Cardiac pacemaker [13, 31]
(Na, K)NbO_3	Piezoelectric, biocompatible	Cardio-mechanical electric sensor [32]
$BaTiO_3$-based	High piezoelectric coefficient, biocompatible	Implantable medical devices [10, 32]
ZnO	Piezoelectric, low toxicity, biodegradable, transparency	Biosensors [33]
PVDF	Piezoelectric, flexibility, elasticity, biocompatibility	In vivo biomechanical energy harvesting [8, 27] Cochlear implant [34]
PVDF-TrFE	Piezoelectric, flexibility, biocompatibility	Energy harvesting [9, 20–24] Pressure sensor [26] Bone tissue engineering [35]
Poly-L-lactic acid	Piezoelectric (shear piezoelectricity in particular), shape-adaptable, biodegradable, biocompatible	Energy harvesting [36] Biodegradable implants [37]

Table 1.
Piezoelectric materials can be used for nanogenerators in biomedical field.

There are many nanogenerator materials that have been reported thus far. Some representative piezoelectric materials that can be used for nanogenerators in biomedical applications are shown in **Table 1**.

In the future, it will be critical to further develop nanogenerator materials with more precise conformity to medical principles and the requirements of clinical applications.

3. Examples of nanogenerator applications in the biomedical field

Energy harvesting systems based on irregular body motions or mechanical deformation are promising candidates for self-powered biomedical devices [1–4]. Using nanogenerators inside the human body is of great medical interest because they can scavenge inexhaustible biomechanical energy from muscle contraction/relaxation, blood circulation, respiration, and cardiac motion and convert it into electrical energy [5–10]. Several examples of the applications of nanogenerators in the biomedical field are illustrated in the following text.

3.1 Nanogenerators can be used as self-powered pressure sensors

Human healthcare monitoring is becoming increasingly significant because of the need for early disease diagnosis and daily health assessments. Conventional monitoring systems are usually powered by batteries, which have limited lifespan and can cause many problems [38]. Self-powered nanogenerators can solve the power supply issue and can be easily integrated into the healthcare system [38]. Some examples are listed below.

A cardiac sensor, used for heart-rate monitoring, is a critical component in personal healthcare management. Self-powered nanogenerators have been employed in self-powered cardiac sensors, as shown in **Figure 1** [5]. Besides the merit of self-powered, they are non-invasive, cost-effective and user-friendly. These implantable cardiac sensors can detect a number of arrhythmic symptoms and provide real-time feedback spontaneously [5]. Compared to current wearable heartbeat monitoring systems, the implantable cardiac sensors can provide both higher accuracy and greater reliability [39]. Self-powered wireless cardiac sensors have a great potential in the future heart healthcare monitoring market.

Physiological parameters such as respiration rate, blood pressure, and pulse rate are major concerns in clinical practice [40]. Failure to detect these signals timely can result in life-threatening conditions [40]. Scientists recently fabricated self-powered TENG-based pressure sensors with a high sensitivity of 150 mV/Pa [41]. When the flexible pressure sensor was attached to the human body, respiration and pulse were accurately and spontaneously monitored [41]. The sensitivity, flexibility, and robustness of nanogenerators allow them to be used in wearable and wrist-based pulse wave detectors [40–43].

3.2 Nanogenerators can be used in pacemakers

When a heart's natural pacemaker is not working properly, resulting in a heart-rate that may be too fast, too slow, or irregular, a doctor may implant a device called pacemaker to restore the heart's nature rhythm. Implantable battery-powered pacemakers, which use electrical impulses to stimulate the heart muscles and regulate heartbeat, have been in clinical use for more than 50 years [13, 15]. Pacemakers have made significant contributions to the treatment of heart diseases such as sick sinus syndrome, heart blockage, and abnormal heart rate [13, 15]. However, every 7–10 years, surgery is needed to replace the pacemaker battery [44, 45]. Self-powered devices can prolong the pacemaker's operation and eliminate battery replacement surgery.

Both PENG and TENG have been investigated for cardiac pacemakers [46, 47]. Generally, PENG are more robust and durable, but their outputs are relatively low.

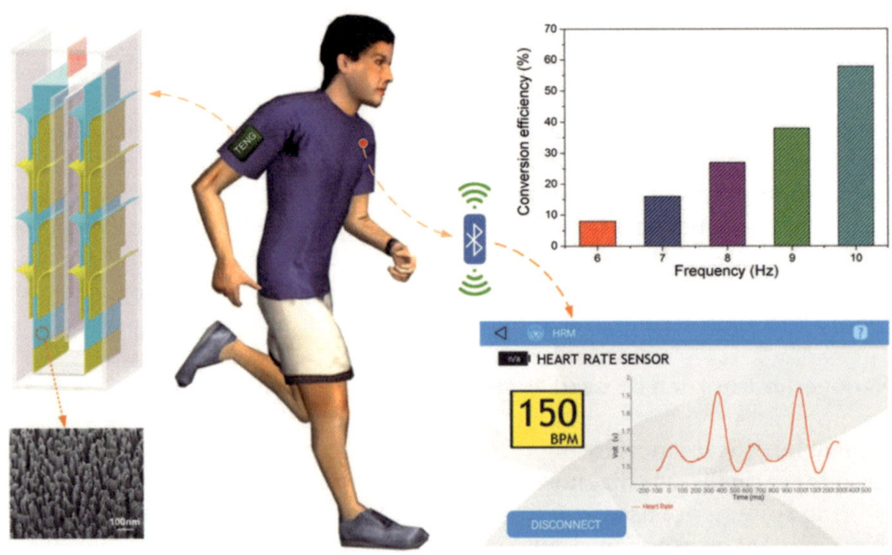

Figure 1.
Illustration of heart-rate monitoring by a wireless self-powered cardiac sensor. Reprinted with permission from [5]. Copyright (2017) American Chemical Society.

TENG materials show a higher output, but they need to be well encapsulated to prevent leakage. A schematic diagram of cardiac pacemaker without battery that can pace the porcine heart is shown in **Figure 2** [31].

3.3 Nanogenerators can be used in cochlear implants

Cochlear implants are neural prosthetic devices that can restore a sense of hearing to people with hearing disability. Cochlear implants work by picking up sound using a microphone located externally above the pinna, and with an external processor, convert the microphone output into electrical pulses that are transmitted internally using a transmitter or receiver to finally stimulate the auditory neurons using an array of electrodes implanted in the cochlea [34]. The conceptual schematics of the cochlear and the basilar membrane are shown in **Figure 3**. However, current cochlear implants have limitations, because they require external components, which are inconvenient for patients. A totally implantable cochlear implant powered by a nanogenerator would address this issue [48]. Scientists have reported the fabrication and characterization of a prototype polyvinylidene

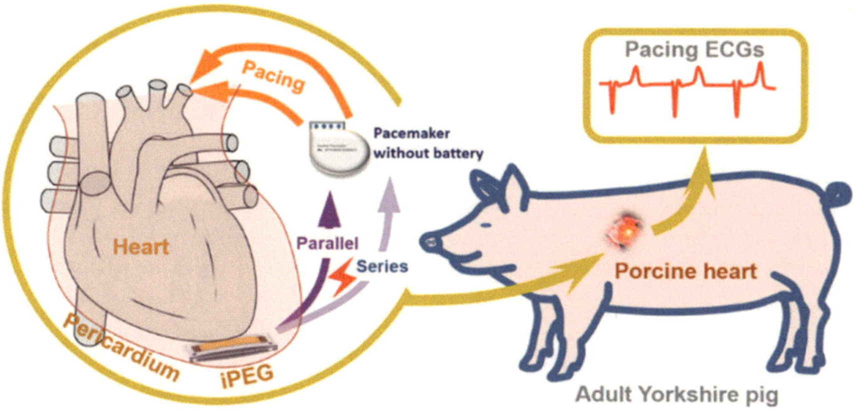

Figure 2.
Schematic diagram of self-powered cardiac pacemaker that pace the porcine heart in vivo. Reprinted with permission from [31]. Copyright (2019) American Chemical Society.

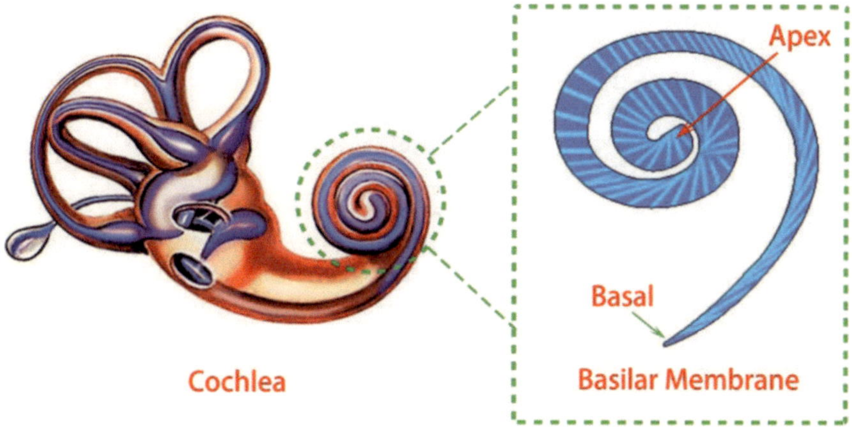

Figure 3.
Conceptual schematics of the cochlear and the basilar membrane. Reprinted with permission from [48]. Copyright © Yudong Liu, et al. 2018.

Cells and tissues	Required electrical fields	Reference
Migration of nerve cells	7 mV/mm	[49]
Migration of embryonic cells	150 mV/mm	[50]
Migration of neural crest cell	150 mV/mm	[51]
Migration of human keratinocytes	10–100 mV/mm	[52]
Wound healing	40–180 mV/ mm	[53, 54]
Cultivation of human bone marrow mesenchymal stem cells	10–600 mV/mm	[55]
Proliferation of osteoblastic cells	20 mV/cm	[56]
Proliferation, migration, and differentiation of muscle precursor cells	Several to tens of nA/cm^2	[57]
Cardiac adipose tissue-derived progenitor cells	5 mV/mm	[58]
Muscle stimulation	mA-level current	[59]

Table 2.
Required electrical fields for cellular and tissue behaviors.

fluoride polymer-based implantable microphone for detecting sound inside gerbil and human cochleae [34]. These results demonstrate the feasibility of the prototype devices as implantable microphones for the development of completely implantable cochlear implants. For patients, this will improve sound reception by utilizing the outer ear and will improve the use of cochlear implants. It should be noted that the development of nanogenerators in cochlear implants field is at the very early stage. They will need further design and innovation to achieve miniaturization, low-power electronics, and an implantable microphone, before they meet the requirements of clinical applications.

3.4 Nanogenerators as stimulators for cells and tissues

Electrical signals play an instructive role in many cellular behaviors, including cell proliferation, differentiation and migration, and tissue wound healing and regeneration. Several examples and their required electrical fields are shown in **Table 2**.

Nanogenerators can provide electrical stimulation for cells and tissues [60–63]. A recent report shows that a self-powered well-aligned P(VDF-TrFE) piezoelectric nanofiber nanogenerator can be used as a piezoelectric stimulator for bone tissue engineering, as shown in **Figure 4** [35]. The well-aligned piezoelectric P(VDF-TrFE) nanogenerators encouraged the MC3T3 cells to proliferate in vitro under a sustainable piezoelectric stimulus. This provides insights into the application of P(VDF-TrFE) piezoelectric nanofiber nanogenerators as a self-powered electrical stimulation system to assist tissue repair and regeneration.

Electrical muscle stimulation is clinically employed for rehabilitative and therapeutic purposes [60]. **Figure 5** illustrates recent research using a stacked-layer triboelectric nanogenerator (TENG) through a flexible multiple-channel intramuscular electrode, which permitted electrical muscle stimulation [60]. Such a self-powered system could be potentially used for rehabilitative and therapeutic purposes to treat muscle function loss.

Nanogenerators have also been developed for skin wound healing. Scientists reported an efficient electrical bandage for accelerated skin wound healing [61]. From in vitro studies, they showed that accelerated skin wound healing could

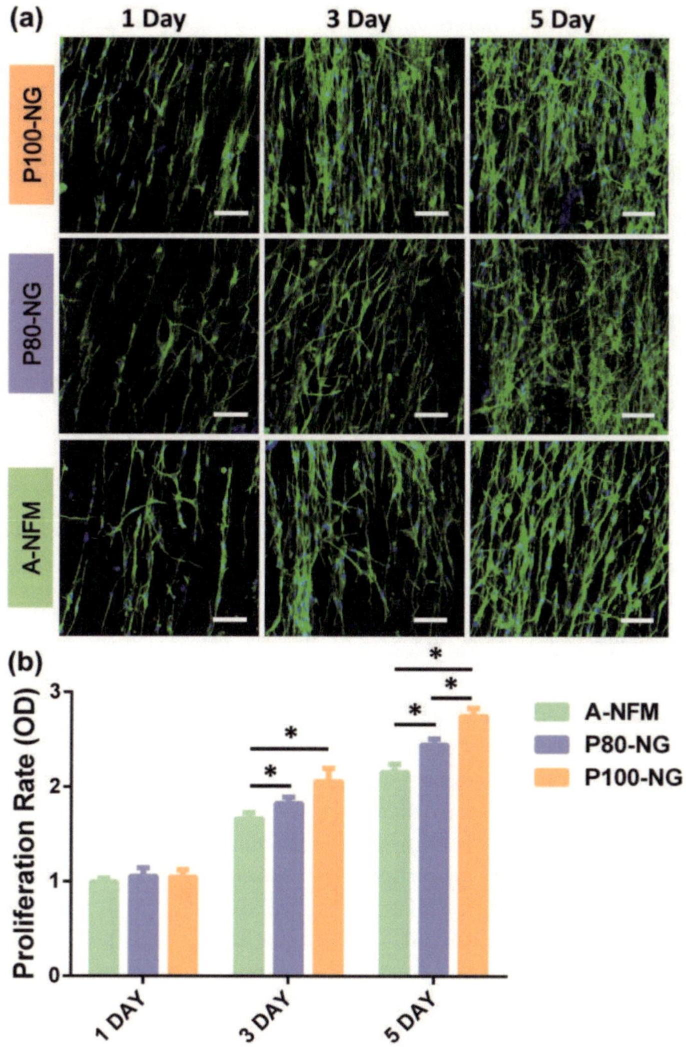

Figure 4.
Proliferation of MC3T3 cells on P80-NG, P100-NG, and control A-NFM. (a) Fluorescence microscopy images of MC3T3 cells on A-NFM, P80-NG, and P100-NG. (b) MC3T3 cells proliferation after 1, 3, and 5 days of culture. The scale bar is 100 μm. The P(VDF-TrFE) nanofiber membranes (NFMs) poled with the electric field of 80 MV/m and 100 MV/m were labeled as P80-NG and P100-NG, respectively. The samples treated by annealing were coded as A-NFM. Reprinted with permission from [35]. Copyright, © 1996–2019 MDPI.

be attributed to electric field-facilitated fibroblast migration, proliferation, and transdifferentiation [61]. This research could lead to a facile therapeutic strategy for nonhealing skin wound treatment.

3.5 Nanogenerators can be used in deep brain stimulators and neural stimulators

Deep brain stimulation is an effective treatment for a variety of neurological disorders, including Parkinson's disease, essential tremor, and epilepsy [64–66]. At present, it involves administering a train of pulses with constant frequency via electrodes implanted in the brain [67]. However, the implantable brain stimulator requires surgery to replace the battery every 3 to 5 years [68]. Self-powered deep brain stimulation is a future technology which does not need external power supply.

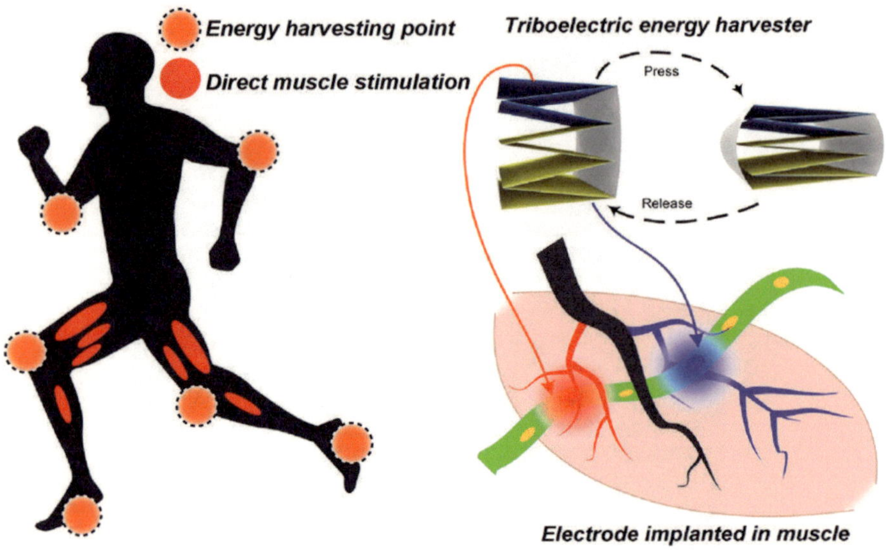

Figure 5.
Schematic illustration of direct electrical muscle stimulation powered by a triboelectric nanogenerator. Reprinted with permission from [60]. Copyright (2019) American Chemical Society.

Scientists have developed a flexible $Pb(In_{1/2}Nb_{1/2})O_3$–$Pb(Mg_{1/3}Nb_{2/3})O_3$–$PbTiO_3$ (PIMNT) energy harvester that can be used in a self-powered deep brain stimulator [68]. More researches in this field open a new avenue for future deep brain stimulation using self-powered deep brain stimulator.

Modulation of neural signals using implantable bioelectronics is an emerging field in fields such as neuroprosthesis and bioelectronic medicine [69–72]. Triboelectric nanogenerators (TENGs) show a promising performance as a power source for neuro-stimulators. Recently, scientists have developed a novel water/air-hybrid TENG that can be used for force-controlled direct stimulation [69]. In another research, scientists present an implanted vagus nerve stimulation system that is battery-free and can spontaneously respond to stomach movement [70]. These provide a concept in therapeutic technology using artificial nerve signal generated from coordinated body activities.

3.6 Nanogenerators as biodegradable electronics

Biodegradable electronics are quite a new scientific term but also an emerging area of research. The general goal is to create human-friendly electronics and enable the integration of electronic circuits with living tissue [73]. Biodegradable electronics, also called transient electronics, are built with degradable organic and inorganic materials, so that they can be integrated with living tissue and used for diagnostic and/or therapeutic purposes during certain physiological processes [74–77]. Once the therapeutic or diagnostic process is completed, the transient devices can be left behind in the body and will degrade and be absorbed gradually without any residue.

Reports show that a biodegradable triboelectric nanogenerator can degrade and be absorbed by the human body after completing its work cycle, so no operation is needed to remove them, leaving no long-term effects [76, 77]. This demonstrates the potential of nanogenerators as a power source for transient medical devices.

Scientists have recently introduced a fully biodegradable nanogenerator based on gelatin film and electro-spun polylactic acid nanofiber membrane, which is fully biodegradable in water [75]. The TENG are disposable and do not harm or pollute the environment.

In general, biodegradable triboelectric nanogenerators offer a promising green micro-power source for biomedical implants, by harvesting energy from body movements, and then dissolve with no adverse effect. The biodegradable medical device field is an emerging area, which shows a great potential for in vivo sensors and therapeutic devices.

4. Future development

The development of nanotechnologies can greatly advance healthcare systems. Nanogenerators can provide complementary or alternative power to traditional batteries in healthcare electronics. Autonomous biomedical devices might be realized with the development of nanogenerators, which will revolutionize the biomedical device and healthcare systems. We expect that autonomous self-powered biomedical systems with active sensing properties are the future development direction of medical devices.

Currently, the key challenges that need to be solved in the field of self-powered implantable medical devices are miniaturization, encapsulation, and stability. There is a strong demand for implantable medical devices with reduced size and weight, to minimize impact on daily activities and increase patient comfort. Also, TENG performance is greatly affected if moisture or liquid leaks into the device when applied in vivo. To avoid corrosion by body fluids, it will be necessary to develop durable and flexible encapsulation to protect the stability and working efficiency of TENG [5, 6].

Future nanogenerator developments in this field are expected to address the following three aspects. Firstly, output performance and energy conversion efficiency should be increased to meet clinical requirements. Secondly, to be used in the human body, nanogenerators need to be highly flexible, sensitive, and durable. For example, many in vivo movements are gentle, and their amplitude is very small, so the nanogenerator must be sensitive enough to exploit small scale motion [7, 14]. Thirdly, since the in vivo environment can be very complex and challenging, careful packaging is needed using biocompatible and soft materials.

In general, nanogenerators have many advantages, including high efficiency, low cost, light weight, and easy fabrication. Nanogenerators have an excellent potential for application in a variety of uses, to provide a sustainable power source for self-powered biomedical electronics and healthcare monitoring systems. With further cutting-edge research and development in this field, a revolution in biomedical devices and healthcare system will be realized in the future.

5. Conclusion

In this chapter, we introduced typical nanogenerator materials that have been developed for biomedical applications. We summarized several examples of how nanogenerators can be used in the biomedical field. We included recent research on nanogenerators in self-powered pressure sensors; pacemakers; cochlear implants; stimulators for cells, tissues, and the brain; and biodegradable electronics. We also pointed out the challenges facing current research and future research directions for nanogenerators in medical devices. We hope this work provides insights and inspiration for future biomedical device and nanogenerator research.

Acknowledgements

This research was supported by the Korea Research Fellowship Program funded by the National Research Foundation of Korea (no. 2017H1D3A1A01054478). This research was also supported by the KUSTAR-KAIST institute, KAIST.

Conflict of interest

The authors declare no conflict of interest.

Author details

Panpan Li, Jeongjae Ryu and Seungbum Hong*
Department of Materials Science and Engineering, KAIST, Daejeon, Korea

*Address all correspondence to: seungbum@kaist.ac.kr

References

[1] Zheng Q, Shi B, Li Z, Wang ZL. Recent progress on piezoelectric and triboelectric energy harvesters in biomedical systems. Advanced Science. 2017;**4**:1700029. DOI: 10.1002/advs.201700029

[2] Feng H, Zhao C, Tan P, Liu R, Chen X, Li Z. Nanogenerator for biomedical applications. Advanced Healthcare Materials. 2018;7:1701298. DOI: 10.1002/adhm.201701298

[3] Proto A, Penhaker M, Conforto S, Schmid M. Nanogenerators for human body energy harvesting. Trends in Biotechnology. 2017;**35**:610-624. DOI: 10.1016/j.tibtech.2017.04.005

[4] Dagdeviren C, Yang BD, Su Y, Tran PL, Joe P, Anderson E, et al. Conformal piezoelectric energy harvesting and storage from motions of the heart, lung, and diaphragm. PNAS. 2014;**111**:1927-1932. DOI: 10.1073/pnas.1317233111

[5] Lin Z, Chen J, Li X, Zhou Z, Meng K, Wei W, et al. Triboelectric nanogenerator enabled body sensor network for self-powered human heart-rate monitoring. ACS Nano. 2017;**11**:8830-8837. DOI: 10.1021/acsnano.7b02975

[6] Zheng Q, Zhang H, Shi B, Xue X, Liu Z, Jin Y, et al. In vivo self-powered wireless cardiac monitoring via implantable triboelectric nanogenerator. ACS Nano. 2016;**10**:6510-6518. DOI: 10.1021/acsnano.6b02693

[7] Liu Z, Zhang S, Jin YM, Ouyang H, Zou Y, Wang XX, et al. Flexible piezoelectric nanogenerator in wearable self-powered active sensor for respiration and healthcare monitoring. Semiconductor Science and Technology. 2017;**32**:064004. DOI: 10.1088/1361-6641/aa68d1

[8] Yu Y, Sun H, Orbay H, Chen F, England CG, Cai W, et al. Biocompatibility and in vivo operation of implantable mesoporous PVDF-based nanogenerators. Nano Energy. 2016;**27**:275-281. DOI: 10.1016/j.nanoen

[9] Kim S, Towfeeq I, Dong Y, Gorman S, Rao AM, Koley G. P(VDF-TrFE) film on PDMS substrate for energy harvesting applications. Applied Sciences. 2018;**8**:213. DOI: 10.3390/app8020213

[10] Yuan M, Cheng L, Xu Q, Wu W, Bai S, Gu L, et al. Biocompatible nanogenerators through high piezoelectric coefficient $0.5Ba(Zr_{0.2}Ti_{0.8})O_3$-$0.5(Ba_{0.7}Ca_{0.3})TiO_3$ nanowires for in-vivo applications. Advanced Materials. 2014;**26**:7432-7437. DOI: 10.1002/adma.201402868

[11] Wang ZL, Song J. Piezoelectric nanogenerators based on zinc oxide nanowire arrays. Science. 2006;**312**:242-246. DOI: 10.1126/science.1124005

[12] Kwon J, Seung W, Sharma BK, Kim SW, Ahn JH. A high performance PZT ribbon-based nanogenerator using graphene transparent electrodes. Energy & Environmental Science. 2012;**5**:8970-8975. DOI: 10.1039/C2EE22251E

[13] Hwang GT, Park H, Lee JH, Oh SK, Park K, Byun M, et al. Self-powered cardiac pacemaker enabled by flexible single crystalline PMN-PT piezoelectric energy harvester. Advanced Materials. 2014;**26**:4880-4887. DOI: 10.1002/adma.201400562

[14] Wang ZL. Triboelectric nanogenerators as new energy technology and self-powered sensors – Principles, problems and perspectives. Faraday Discussions. 2014;**176**:447. DOI: 10.1039/c4fd00159a

[15] Zheng Q, Shi B, Fan F, Wang X, Yan L, Yuan W, et al. In vivo powering of pacemaker by breathing-driven implanted triboelectric nanogenerator.

Advanced Materials. 2014;**26**:5851-5856. DOI: 10.1002/adma.201402064

[16] Zebda A, Alcaraz JP, Vadgama P, Shleev S, Minteer SD, Boucher F, et al. Challenges for successful implantation of biofuel cells. Bioelectrochemistry. 2018;**124**:57-72. DOI: 10.1016/j.bioelechem.2018.05.011

[17] Katz E, MacVittie K. Implanted biofuel cells operating in vivo – methods, applications and perspectives. Energy & Environmental Science. 2013;**6**:2791-2803. DOI: 10.1088/1742-6596/476/1/012063

[18] Núñez CG, Manjakkal L, Dahiya R. Energy autonomous electronic skin. npj Flexible Electronics. 2019;**3**:1. DOI: 10.1038/s41528-018-0045-x

[19] Núñez CG, Navaraj WT, Polat EO, Dahiya R. Energy-autonomous, flexible, and transparent tactile skin. Advanced Functional Materials. 2017;**27**:1606287. DOI: 10.1002/adfm.201606287

[20] Kim D, Hong S, Li D, Roh HS, Ahn G, Kim J, et al. A spring-type piezoelectric energy harvester. RSC Advances. 2013;**3**:3194. DOI: 10.1039/C2RA22554A

[21] Li DJ, Hong S, Gu S, Choi YY, Nakhmanson S, Heinonen O, et al. Polymer piezoelectric energy harvesters for low wind speed. Applied Physics Letters. 2014;**104**:012902. DOI: 10.1063/1.4861187

[22] Kim D, Roh HS, Kim Y, No K, Hong S. Selective current collecting design for spring-type energy harvesters. RSC Advances. 2015;**5**:10662-10666. DOI: 10.1039/C4RA16443A

[23] Ryu J, No K, Kim Y, Park E, Hong S. Synthesis and application of ferroelectric poly(Vinylidene fluoride-co-trifluoroethylene) films using electrophoretic deposition. Scientific Reports. 2016;**6**:36176. DOI: 10.1038/srep36176

[24] Ryu J, Kim J, Oh J, Lim S, Sim JY, Jeon JS, et al. Intrinsically stretchable multi-functional fiber with energy harvesting and strain sensing capability. Nano Energy. 2019;**55**:348-353. DOI: 10.1016/j.nanoen.2018.10.071

[25] Whiter RA, Narayan V, Kar-Narayan S. A scalable nanogenerator based on self-poled piezoelectric polymer nanowires with high energy conversion efficiency. Advanced Energy Materials. 2014;**4**:1400519. DOI: 10.1002/aenm.201400519

[26] Lee JH, Yoon HJ, Kim TY, Gupta MK, Lee JH, Seung W, et al. Micropatterned P(VDF-TrFE) film-based piezoelectric nanogenerators for highly sensitive self-powered pressure sensors. Advanced Functional Materials. 2015;**25**:3203-3209. DOI: 10.1002/adfm.201500856

[27] Sun C, Shi J, Bayerl DJ, Wang X. PVDF microbelts for harvesting energy from respiration. Energy & Environmental Science. 2011;**4**:4508. DOI: 10.1039/c1ee02241e

[28] Park KI, Son JH, Hwang GT, Jeong CK, Ryu J, Koo M, et al. Highly efficient, flexible piezoelectric PZT thin film nanogenerator on plastic substrates. Advanced Materials. 2014;**26**:2154. DOI: 10.1002/adma.201305659

[29] Luo C, Omelchenko I, Manson R, Robbins C, Oesterle EC, Cao GZ, et al. Direct intracochlear acoustic stimulation using a PZT microactuator. Trends in Hearing. 2015;**19**:2331216515616942. DOI: 10.1177/2331216515616942

[30] Lu C, Wu S, Lu B, Zhang Y, Du Y, Feng X. Ultrathin flexible piezoelectric sensors for monitoring eye fatigue. Journal of Micromechanics and

Microengineering. 2018;**28**:025010. DOI: 10.1088/1361-6439/aaa219

[31] Li N, Yi Z, Ma Y, Xie F, Huang Y, Tian Y, et al. Direct powering a real cardiac pacemaker by natural energy of a heartbeat. ACS Nano. 2019;**13**:2822. DOI: 10.1021/acsnano.8b08567

[32] Kwak J, Kingon A, Kim SH. Lead-free $(Na_{0.5},K_{0.5})NbO_3$ thin films for the implantable piezoelectric medicalsensor applications. Materials Letters. 2012;**82**:130-132. DOI: 10.1016/j.matlet.2012.05.079

[33] Zhu P, Weng Z, Li X, Liu X, Wu S, Yeung KWK, et al. Biomedical applications of functionalized ZnO nanomaterials: From biosensors to bioimaging. Advanced Materials Interfaces. 2016;**3**:1500494. DOI: 10.1002/admi.201500494

[34] Park S, Guan X, Kim Y, Creighton FX, Wei E, Kymissis I, et al. PVDF-based piezoelectric microphone for sound detection inside the cochlea: Toward totally implantable cochlear implants. Trends in Hearing. 2018;**22**:2331216518774450. DOI: 10.1177/2331216518774450

[35] Wang A, Hu M, Zhou L, Qiang X. Self-powered well-aligned P(VDF-TrFE) piezoelectric nanofiber nanogenerator for modulating an exact electrical stimulation and enhancing the proliferation of preosteoblasts. Nanomaterials. 2019;**9**:349. DOI: 10.3390/nano9030349

[36] Zhu J, Jia L, Huang R. Electrospinning poly(l-lactic acid) piezoelectric ordered porous nanofibers for strain sensing and energy harvesting. Journal of Materials Science: Materials in Electronics. 2017;**28**:12080. DOI: 10.1007/s10854-017-7020-5

[37] Pan R, Xuan W, Chen J, Dong S, Jin H, Wang X, et al. Fully biodegradable triboelectric nanogenerators based on electrospun polylactic acid and nanostructured gelatin films. Nano Energy. 2018;**45**:193. DOI: 10.1016/j.nanoen.2017.12.048

[38] Song P, Yang G, Lang T, Yong KT. Nanogenerators for wearable bioelectronics and biodevices. Journal of Physics D: Applied Physics. 2019;**52**:023002. DOI: 10.1088/1361-6463/aae44d

[39] Krahn AD, Pickett RA, Sakaguchi S, Shaik N, Cao J, Norman HS, et al. R-wave sensing in an implantable cardiac monitor without ECG-based preimplant mapping: Results from a multicenter clinical trial. Pacing and Clinical Electrophysiology. 2014;**37**:505. DOI: 10.1111/pace.12303

[40] Ma Y, Zheng Q, Liu Y, Shi B, Xue X, Ji W, et al. Self-powered, one-stop, and multifunctional implantable triboelectric active sensor for real-time biomedical monitoring. Nano Letters. 2016;**16**:6042. DOI: 10.1021/acs.nanolett.6b01968

[41] Liu Z, Zhao Z, Zeng X, Fu X, Hu Y. Expandable microsphere-based triboelectric nanogenerators as ultrasensitive pressure sensors for respiratory and pulse monitoring. Nano Energy. 2019;**59**:295-301. DOI: 10.1016/j.nanoen.2019.02.057

[42] Lin MF, Xiong J, Wang J, Parida K, Lee PS. Core-shell nanofiber mats for tactile pressure sensor and nanogenerator applications. Nano Energy. 2018;**44**:248-255. DOI: 10.1016/j.nanoen.2017.12.004

[43] Jin L, Tao J, Bao R, Sun L, Pan C. Self-powered real-time movement monitoring sensor using triboelectric nanogenerator technology. Scientific Reports. 2017;**7**:10521. DOI: 10.1038/s41598-017-10990-y

[44] DiFrancesco D. Pacemaker mechanisms in cardiac tissue. Annual

Review of Physiology. 2013;**55**:455-472. DOI: 10.1146/annurev.ph.55.030193.002323

[45] Hu YF, Lin L, Zhang Y, Wang ZL. Replacing a battery by a nanogenerator with 20 V output. Advanced Materials. 2012;**24**:110. DOI: 10.1002/adma.201103727

[46] Choi MY, Choi D, Jin MJ, Kim I, Kim SH, Choi JY, et al. Mechanically powered transparent flexible charge generating nanodevices with piezoelectric ZnO nanorods. Advanced Materials. 2009;**21**:2185. DOI: 10.1002/adma.200803605

[47] Park KI, Xu S, Liu Y, Hwang GT, Kang SJ, Wang ZL, et al. Piezoelectric $BaTiO_3$ thin film nanogenerator on plastic substrates. Nano Letters. 2010;**10**:4939. DOI: 10.1021/nl102959k

[48] Liu Y, Zhu Y, Liu J, Zhang Y, Liu J, Zhai J. Design of bionic cochlear basilar membrane acoustic sensor for frequency selectivity based on film triboelectric nanogenerator. Nanoscale Research Letters. 2018;**13**:191. DOI: 10.1186/s11671-018-2593-3

[49] Robinson KR. The responses of cells to electrical fields: A review. The Journal of Cell Biology. 1985;**101**:2023-2027. DOI: 10.1083/jcb.101.6.2023

[50] Nuccitelli R, Erickson CA. Embryonic cell mobility can be guided by physiological electric fields. Experimental Cell Research. 1983;**147**:195-201

[51] Gruler H, Nuccitelli R. Neural crest cell galvanotaxis: New data and a novel approach to the analysis of both galvanotaxis and chemotaxis. Cell Motility and the Cytoskeleton. 1991;**19**:121-133. DOI: 10.1002/cm.970190207

[52] Nishimura KY, Isseroff RR, Nuccitelli R. Human keratinocytes migrate to the negative pole in direct current electric fields comparable to those measured in mammalian wounds. Journal of Cell Science. 1996;**109**:199-207

[53] Zhao M, Song B, Pu J, Wada T, Reid B, Tai G, et al. Electrical signals control wound healing through phosphatidylinositol-3-OH kinase-γ and PTEN. Nature. 2006;**442**:457-460. DOI: 10.1038/nature04925

[54] Zhao M. Electrical fields in wound healing—An overriding signal that directs cell migration. Seminars in Cell & Developmental Biology. 2009;**20**:674-682. DOI: 10.1016/j.semcdb.2008.12.009

[55] Zhao Z, Watt C, Karystinou A, Roelofs AJ, McCaig CD, Gibson IR, et al. Directed migration of human bone marrow mesenchymal stem cells in a physiological direct current electric field. European Cells & Materials. 2011;**22**:344-358. DOI: 10.22203/eCM.v022a26

[56] Zhuang H, Wang W, Seldes RM, Tahernia AD, Fan H, Brighton CT. Electrical stimulation induces the level of TGF-b1 mRNA in osteoblastic cells by a mechanism involving calcium/calmodulin pathway. Biochemical and Biophysical Research Communications. 1997;**237**:225-229

[57] Lee JH, Jeon WY, Kim HH, Lee EJ, Kim HW. Electrical stimulation by enzymatic biofuel cell to promote proliferation, migration and differentiation of muscle precursor cells. Biomaterials. 2015;**53**:358-369. DOI: 10.1016/j.biomaterials.2015.02.062

[58] Llucià-Valldeperas A, Sanchez B, Soler-Botija C, Gálvez-Montón C, Prat-Vidal C, Roura S, et al. Electrical stimulation of cardiac adipose tissue-derived progenitor cells modulates cell phenotype and genetic machinery. Journal of Tissue Engineering and Regenerative Medicine. 2015;**9**:E76-E83. DOI: 10.1002/term.1710

[59] Peckham PH, Knutson JS. Functional electrical stimulation for neuromuscular applications. Annual Review of Biomedical Engineering. 2005;**7**:327-360. DOI: 10.1146/annurev.bioeng.6.040803.140103

[60] Wang J, Wang H, Thakor NV, Lee C. Self-powered direct muscle stimulation using a triboelectric nanogenerator (TENG) integrated with a flexible multiple-channel intramuscular electrode. ACS Nano. 2019;**13**:3589-3599. DOI: 10.1021/acsnano.9b00140

[61] Long Y, Wei H, Li J, Yao G, Yu B, Ni D, et al. Effective would healing enabled by discrete alternative electric fields from wearable nanogenerators. ACS Nano. 2018;**12**:12533-12540. DOI: 10.1021/acsnano.8b07038

[62] Murillo G, Blanquer A, Vargas-Estevez C, Barrios L, Ibáñez E, Nogués C, et al. Electromechanical nanogenerator–cell interaction modulates cell activity. Advanced Materials. 2017;**29**:1605048. DOI: 10.1002/adma.201605048

[63] Hu W, Wei X, Zhu L, Yin D, Wei A, Bi X, et al. Enhancing proliferation and migration of fibroblast cells by electric stimulation based on triboelectric nanogenerator. Nano Energy. 2019;**57**:600. DOI: 10.1016/j.nanoen.2018.12.077

[64] Groiss SJ, Wojtecki L, Südmeyer M, Schnitzler A. Deep brain stimulation in Parkinson's disease. Therapeutic Advances in Neurological Disorders. 2009;**2**:20. DOI: 10.1177/1756285609339382

[65] Flora ED, Perera CL, Cameron AL, Maddern GJ. Deep brain stimulation for essential tremor: A systematic review. Movement Disorders. 2010;**25**:1550. DOI: 10.1002/mds.23195

[66] Salanova V. Deep brain stimulation for epilepsy. Epilepsy & Behavior. 2018;**88S**:21-24. DOI: 10.1016/j.yebeh.2018.06.041

[67] Weerasinghe G, Duchet B, Cagnan H, Brown P, Bick C, Bogacz R. Predicting the effects of deep brain stimulation using a reduced coupled oscillator model. PLoS Computational Biology. 2019;**15**:e1006575. DOI: 10.1371/journal.pcbi.1006575

[68] Hwang GT, Kim Y, Lee JH, Oh S, Jeong CK, Park DY, et al. Self-powered deep brain stimulation via a flexible PIMNT energy harvester. Energy & Environmental Science. 2015;**8**:2677-2684. DOI: 10.1039/c5ee01593f

[69] Lee S, Wang H, Wang J, Shi Q, Yen SC, Thakor NV, et al. Battery-free neuromodulator for peripheral nerve direct stimulation. Nano Energy. 2018;**50**:148. DOI: 10.1016/j.nanoen.2018.04.004

[70] Yao G, Kang L, Li J, Long Y, Wei H, Ferreira CA, et al. Effective weight control via an implanted self-powered vagus nerve stimulation device. Nature Communications. 2018;**9**:5349. DOI: 10.1038/s41467-018-07764-z

[71] Famm K, Litt B, Tracey KJ, Boyden ES, Slaoui M. A jump-start for electroceuticals. Nature. 2013;**496**:159. DOI: 10.1038/496159a

[72] Adameyko I. Neural circuitry gets rewired. Science. 2016;**354**:833. DOI: 10.1126/science.aal2810

[73] Irimia-Vladu M. "Green" electronics: Biodegradable and biocompatible materials and devices for sustainable future. Chemical Society Reviews. 2014;**43**:588. DOI: 10.1039/c3cs60235d

[74] Kim DH, Viventi J, Amsden JJ, Xiao J, Vigeland L, Kim YS, et al. Dissolvable films of silk fibroin for ultrathin conformal bio-integrated electronics. Nature Materials. 2010;**9**:511. DOI: 10.1038/nmat2745

[75] Pan R, Xuan W, Chen J, Dong S, Jin H, Wang X, et al. Fully biodegradable triboelectric nanogenerators based on electro-spun polylactic acid and nanostructured gelatin films. Nano Energy. 2018;**45**:193. DOI: 10.1016/j.nanoen.2017.12.048

[76] Zheng Q, Zou Y, Zhang Y, Liu Z, Shi B, Wang X, et al. Biodegradable triboelectric nanogenerator as a life-time designed implantable power source. Science Advances. 2016;**2**:e1501478. DOI: 10.1126/sciadv.1501478

[77] Liang Q, Zhang Q, Yan X, Liao X, Han L, Yi F, et al. Recyclable and green triboelectric nanogenerator. Advanced Materials. 2017;**29**:1604961. DOI: 10.1002/adma.201604961